Recent Advances in Biotechnology
(Volume 3)
Recent Progress in Glycotherapy

Edited by

Qun Zhou
Protein Engineering,
Biologics Research, Sanofi, Framingham,
Massachusetts 01701,
United States

Recent Advances in Biotechnology

Volume # 3

Recent Progress in Glycotherapy

Editor: Qun Zhou

ISSN (Online): 2468-5372

ISSN: Print: 2468-5364

ISBN (eBook): 978-1-68108-391-9

ISBN (Print): 978-1-68108-392-6

advertisements or ideas contained in the Work.

Limitation of Liability:

In no event will Bentham Science Publishers, its staff, editors and/or authors, be liable for any damages, including, without limitation, special, incidental and/or consequential damages and/or damages for lost data and/or profits arising out of (whether directly or indirectly) the use or inability to use the Work. The entire liability of Bentham Science Publishers shall be limited to the amount actually paid by you for the Work.

General:

1. Any dispute or claim arising out of or in connection with this License Agreement or the Work (including non-contractual disputes or claims) will be governed by and construed in accordance with the laws of the U.A.E. as applied in the Emirate of Dubai. Each party agrees that the courts of the Emirate of Dubai shall have exclusive jurisdiction to settle any dispute or claim arising out of or in connection with this License Agreement or the Work (including non-contractual disputes or claims).
2. Your rights under this License Agreement will automatically terminate without notice and without the need for a court order if at any point you breach any terms of this License Agreement. In no event will any delay or failure by Bentham Science Publishers in enforcing your compliance with this License Agreement constitute a waiver of any of its rights.
3. You acknowledge that you have read this License Agreement, and agree to be bound by its terms and conditions. To the extent that any other terms and conditions presented on any website of Bentham Science Publishers conflict with, or are inconsistent with, the terms and conditions set out in this License Agreement, you acknowledge that the terms and conditions set out in this License Agreement shall prevail.

Bentham Science Publishers Ltd.
Executive Suite Y - 2
PO Box 7917, Saif Zone
Sharjah, U.A.E.
Email: subscriptions@benthamscience.org

**BENTHAM
SCIENCE**

CONTENTS

FOREWORD

This book includes a timely collection of topics addressing important advances in glycotechnology that are being exploited in the development of glycotherapeutics.

This area is rooted in the rich history of the chemistry of carbohydrates that contributed to defining the structures of oligosaccharides and developed the complex synthesis of these exquisite molecules. In addressing the functions of these structures in animal cells glycobiologists, using the major advances in biochemistry and molecular biology developed in the last decades of the 20th century, unraveled the complex process involved in the biosynthesis, assembly, and processing of glycoproteins, glycolipids and proteoglycans. One of the seminal observations that initiated the general area of glycobiology was Victor Ginsburg's discovery in the early 1960's that neuraminidase digestion of lymphocytes altered their normal trafficking pattern, which ultimately led to the discovery of the animal cell lectins, and we now know that the functions of mammalian cell glycans are somehow associated with either direct or indirect protein-glycan interactions making them ideal candidates for glycotherapeutic studies with many examples already available.

The obvious challenge now resides in the continued search for the roles of glycans in normal physiology and disease; and these advances a coming rapidly in the field we now refer to as Glycomics, which has recently been recognized as a strategic area for NIH Common Fund support. The future is indeed bright, and the opportunities are ripe for the investigators to exploit this rapidly expanding area for glycotherapy.

Dr. David F. Smith
Emory Comprehensive Glycomics Core
Emory University School of Medicine
Atlanta, GA 30322
USA
E-mail: dfsmith@emory.edu

PREFACE

There are many therapies being developed through biotechnology for treating diseases. They include protein therapy, gene therapy, cell therapy, and glycotherapy. As one of the important disease treatment approaches, glycotherapy provides solution for some unmet medical needs using glycoengineering. It has been used in clinics for decades.

Heparin, a glycosaminoglycan and blood thinner, acts as an anticoagulant and is one of World Health Organization's lists of essential medicines. The sialic acid analogues, oseltamivir and zanamivir, are widely used anti-influenza drugs. Polysaccharide vaccines including Menactra, Prevnar, and Typhim Vi, have been used to stimulate immunity against infection. In addition to therapy using sugar or sugar analogues, therapeutic proteins, such as darbepoetin alfa and imiglucerase, are also generated using glycoengineering *in vivo* or *in vitro* to enhance therapeutic index.

During the last two decades, significant progress has been made in glycotherapy with glycoengineering and glycan mimics although there are many challenges. The glycans are chemoenzymatically modified or conjugated to small molecular weight drugs, proteins, and nucleic acids for increasing pharmacokinetics and pharmacodynamics. Although there is an excellent review published recently by Hudak and Bertozzi (Chemistry & Biology 2014), glycotherapy has not been extensively appreciated in books and unknown to many readers outside this particular field. The current book attempts to fill the gap and provide more recent information related to the exciting progress in this important area of biotechnology.

The book includes reviews on glycotherapy which focus on modification of protein or small molecular weight drug using recombinant and chemoenzymatic approaches. The progress on bioconjugation of glycan using hydrophilic polymers is also covered.

Chapters 1 and 2 focus on the development of vaccines and antibodies against tumor-associated carbohydrate antigen for cancer treatment, as well as the use of glycan for viral inhibition. Chapters 3 and 4 describe the progress in glycoPEGylation and hyaluronic acid conjugation for increasing therapeutic index in treating diseases. The glycoengineering of therapeutic proteins is reviewed in chapters 5, 6 and 7. They provide overviews of recent advances in modification of glycans in proteins or antibodies using recombinant, chemoenzymatic or bioconjugation methods.

Although only a few topics of glycotherapy are being reviewed here, this book aims to provide readers with overview of the researches which have been actively pursued during recent years. We expect to have more excellent books or reviews on this important therapeutic area in the future.

Dr. Qun Zhou
Protein Engineering
Global BioTherapeutics, Sanofi, Framingham
Massachusetts 01701
United States

List of Contributors

Anna Mero — Department of Pharmaceutical and Pharmacological Sciences, University of Padova, Via F. Marzolo 5, 35131 Padova, Italy

Antonella Grigoletto — Department of Pharmaceutical and Pharmacological Sciences, University of Padova, Via F. Marzolo 5, 35131 Padova, Italy

Che C. Colpitts — Inserm, U1110, Institut de Recherche sur les Maladies Virales et Hépatiques, 67000 Strasbourg, France.
Université de Strasbourg, 67000 Strasbourg, France

Diana P. Sousa — UCIBIO, Departamento de Ciências da Vida, Faculdade de Ciências e Tecnologia, Universidade NOVA de Lisboa, Portugal.

Gabriele Martinez — Veneto Institute of Oncology IOV-IRCCS, Padova, Italy

Gianfranco Pasut — Department of Pharmaceutical and Pharmacological Sciences, University of Padova, Via F. Marzolo 5, 35131 Padova, Italy.
Veneto Institute of Oncology IOV-IRCCS, Padova, Italy

Huawei Qiu — Protein Engineering, Sanofi, 5 Mountain Road, Framingham, MA 01701, United States

Huijuan Li — Moderna Therapeutics 200 Technology Square Cambridge, MA 02139, USA

Kevin B. Turner — Sterile Product and Analytical Development, Biologics and Vaccine Development, Merck & Co., Inc., Kenilworth, NJ USA

Liliana R. Loureiro — UCIBIO, Departamento de Ciências da Vida, Faculdade de Ciências e Tecnologia, Universidade NOVA de Lisboa, Portugal.

Marcos Oggero — UNL, CONICET, Cell Culture Laboratory, FBCB. Edificio FBCB-Ciudad Universitaria UNL. C.C. 242. (S3000ZAA) Santa Fe., Argentina

Marina Etcheverrigaray — UNL, CONICET, Cell Culture Laboratory, FBCB. Edificio FBCB-Ciudad Universitaria UNL. C.C. 242. (S3000ZAA) Santa Fe., Argentina

M. Eugenia Giorgi — CIHIDECAR-CONICET, Departamento de Química Orgánica, Facultad de Ciencias Exactas y Naturales, Universidad de Buenos Aires, Buenos Aires, Argentina

Mylène A. Carrascal — UCIBIO, Departamento de Ciências da Vida, Faculdade de Ciências e Tecnologia, Universidade NOVA de Lisboa, Portugal.

Natalia Ceaglio — UNL, CONICET, Cell Culture Laboratory, FBCB. Edificio FBCB-Ciudad Universitaria UNL. C.C. 242. (S3000ZAA) Santa Fe., Argentina

Paula A. Videira — UCIBIO, Departamento de Ciências da Vida, Faculdade de Ciências e Tecnologia, Universidade NOVA de Lisboa, Portugal.
Univ. Lille, Inserm, U908 - CPAC - Cell Plasticity and Cancer, F-59000 Lille, France

Philippe Delannoy	Univ. Lille, CNRS, UMR 8576 - UGSF - Unité de Glycobiologie Structurale et Fonctionnelle, F-59000 Lille, France
Qun Zhou	Protein Engineering, Sanofi, 5 Mountain Road, Framingham, MA 01701, United States
Ricardo Kratje	UNL, CONICET, Cell Culture Laboratory, FBCB. Edificio FBCB-Ciudad Universitaria UNL. C.C. 242. (S3000ZAA) Santa Fe., Argentina
Robert Yite Chou	Sterile Product and Analytical Development, Biologics and Vaccine Development, Merck & Co., Inc., Kenilworth, NJ USA
Rosa M. de Lederkremer	CIHIDECAR-CONICET, Departamento de Química Orgánica, Facultad de Ciencias Exactas y Naturales, Universidad de Buenos Aires, Buenos Aires, Argentina
Rosalia Agusti	CIHIDECAR-CONICET, Departamento de Química Orgánica, Facultad de Ciencias Exactas y Naturales, Universidad de Buenos Aires, Buenos Aires, Argentina
Sylvain Julien	Univ. Lille, Inserm, U908 - CPAC - Cell Plasticity and Cancer, F-59000 Lille, France
Thomas F. Baumert	Inserm, U1110, Institut de Recherche sur les Maladies Virales et Hépatiques, 67000 Strasbourg, France. Université de Strasbourg, 67000 Strasbourg, France. Institut Hospitalo-Universitaire, Pôle Hépato-digestif, Hôpitaux Universitaires de Strasbourg, 67000 Strasbourg, France

Recent Advances in
Biotechnology
(Volume 3)
Recent Progress in Glycotherapy

2

CHAPTER 1

Vaccine and Antibody Therapy Against Thomsen-Friedenreich Tumor-Associated Carbohydrate Antigens

Paula A. Videira[1,2,*], Sylvain Julien[2], Liliana R. Loureiro[1], Diana P. Sousa[1], Mylène A. Carrascal[1] and Philippe Delannoy[3,*]

[1] *UCIBIO, Departamento de Ciências da Vida, Faculdade de Ciências e Tecnologia, Universidade NOVA de Lisboa, Portugal*

[2] *Univ. Lille, Inserm, U908 - CPAC - Cell Plasticity and Cancer, F-59000 Lille, France*

[3] *Univ. Lille, CNRS, UMR 8576 - UGSF - Unité de Glycobiologie Structurale et Fonctionnelle, F-59000 Lille, France*

Abstract: The Thomsen Friedenreich (TF) carbohydrate antigens are a group of short *O*-glycans overexpressed on most carcinomas that have been correlated with cancer progression and poor prognosis. Usually associated with immunosuppressive tumor environment, there is a number of potential immunotherapeutics against TF antigens have been developed, which comprise vaccines and antibodies. As therapeutic vaccination, TF antigens already entered into clinical trials, but with limited success due to low patient's response. Novel vaccine design, with multiantigenicity and pointing towards the cellular immune responses arises as a potent stratagem to overcome the low ability of TF antigens to boost immune responses, typical of carbohydrates. The development of antibodies against TF antigens boosted even before vaccine development. These are mainly used for diagnostics, but so far no such antibody entered into clinical trials in patients. Increasing the specificity and the therapeutic efficiency of existing antibodies and developing novel antibodies are still necessary. The vast array of methodologies and engineering techniques available today will allow rapid development and novel formats for both vaccines and antibodies.

* Corresponding authors Paula Videira: UCIBIO, Departamento de Ciências da Vida, Faculdade de Ciências e Tecnologia, Universidade NOVA de 2829-516 Caparica, Portugal; Tel: +351 968169892; E-mail: p.videira@fct.unl.pt, **Philippe Delannoy:** Université de Lille - Sciences et Technologies Unité de Glycobiologie Structurale et Fonctionnelle (UGSF) UMR CNRS 8576, Bât. C9 - 59655 Villeneuve d'Ascq, France; Phone +33 (0)3 20436923; E-mail: philippe.delannoy@univ-lille1.fr

Vaccines and antibodies targeting the same epitopes can function in synergy both to protect and to clear patient's cancer cells. Whilst on one hand, TF antigens dampen immune responses against tumor cells, it is anticipated that the challenge is overcome by applying our increasing knowledge of the mechanisms behind to improve molecular design. Novel solutions are also envisaged by combining anti-TF therapies with other immunotherapies.

Keywords: Clinical trials, Glycan-based vaccines, IgG, Immune response, Immunological memory, Immunotherapy, Mimetic vaccines, Monoclonal antibodies, Mucins, O-glycans, Self-adjuvanting vaccines, Sialic acid, Sialyl-Tn antigens, T antigen, Therapeutic antibodies, Theratope, Thomsen Friedenreich antigens, Tn antigen, Tumor associated carbohydrates.

INTRODUCTION

Structure and Biosynthesis of Thomsen–Friedenreich Antigens

Thomsen–Friedenreich antigens are *O*-linked carbohydrate antigens found on membrane glycoproteins, especially on serine and threonine rich and tandem repeated domains of mucins [1]. The Thomsen–Friedenreich antigens are originated during the first steps of mucin glycosylation, resulting from a defect in the elongation of *O*-glycan chains in cancer cells.

The first described Thomsen-Friedenreich antigen (called T or TF antigen) was initially found on red blood cells and consists in the disaccharide Galβ1-3GalNAc α-linked to serine or threonine residue (Galβ1-3GalNAcα1-Ser/Thr), forming the core 1 structure of mucin *O*-glycans. T antigen is normally the precursor of core 2 *O*-glycans but can be unmasked when cancer cells have lost the ability to synthesize the core 2 (Fig. **1**). The second described TF antigen called Tn antigen is the cryptic precursor of the core 1 and consists in the single *N*-acetyl-galactosamine (GalNAc) residue α-linked to serine or threonine residue (GalNAcα1-Ser/Thr) that can remain unmasked if the ability to synthesize core 1 is lost (Fig. **1**) [2]. The expression of Tn antigen on blood cells is responsible for the Tn-syndrome, which is a rare autoimmune hematological disorder [3]. The Tn antigen can be also sialylated on C6 position of GalNAc, resulting in the disaccharide Neu5Acα2-6GalNAcα1-Ser/Thr, known as sialyl-Tn (STn) antigen.

STn is almost absent in normal healthy tissues but can be detected at various frequencies in almost all kind of carcinomas [4].

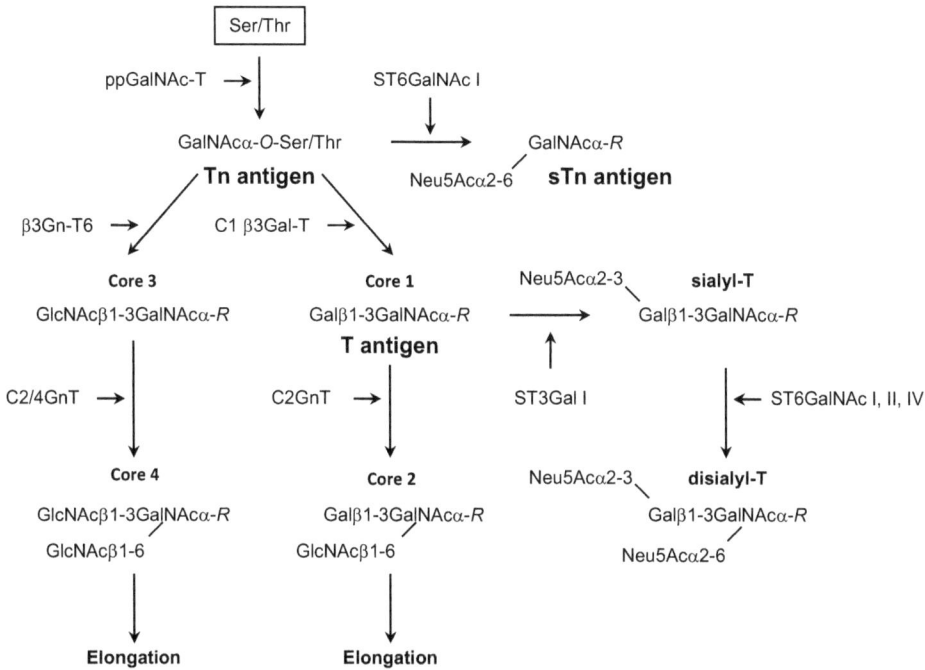

Fig. (1). Biosynthesis of *O*-glycan chains in normal and cancer cells. The figure illustrates mechanisms by which *O*-glycans expressed in normal cells can be turned down to Thomsen–Friedenreich antigens (T, Tn and STn antigens). Gal: Galactose, GalNAc: *N*-acetyl-galactosamine, GlcNAc: *N*-acetyl-glucosamine, Neu5Ac: *N*-acetyl-neuraminic acid, Fuc: Fucose.

The biosynthesis of mucin *O*-glycan chains is a step-by-step process occurring in the Golgi apparatus. *O*-glycans are synthesized by the sequential action of several glycosyltransferases, each transferring a monosaccharide *e.g.*, *N*-acetyl-galactosamine (GalNAc), galactose (Gal), *N*-acetyl-glucosamine (GlcNAc), or *N*-acetyl-neuraminic acid (Neu5Ac) from a donor nucleotide-sugar (*e.g.*, UDP-GalNAc, UDP-Gal, UDP-GlcNAc or CMP-Neu5Ac) to an acceptor that is the glycan being synthesized. Glycosyltransferases are membrane-bound enzymes of which the level of expression, substrate specificity and localization in Golgi compartments are responsible for the pattern of *O*-glycans expressed in a given cell or carried by a given glycoprotein. The initiation step is the transfer of a GalNAc residue on a serine or a threonine of the tandem repeat sequences of apomucins by an extensive family of UDP-GalNAc: polypeptide *N*-acetyl-galactosaminyltransferases (ppGalNAc-Ts) forming the Tn antigen (GalNAcα1-Ser/Thr) (Fig. **1**). Despite the apparent simplicity of ppGalNAc-Ts catalytic function, fifteen members of the family have been functionally characterized in mammals, and *in silico* analysis indicates that as many as 24 ppGalNAc-Ts may exist [5]. ppGalNAc-T isoforms display tissue-specific expression, with some being broadly expressed while others are more restricted to certain cells or tissues. This diversity allows a finely tuned control of the initiation of *O*-glycosylation in a cell-specific and protein-specific manner.

Four main core structures (cores 1 to 4) occur in human mucins (Fig. **1**). The biosynthesis of cores 1 and 2 begins by the transfer of a Gal residue in β1,3-linkage onto Tn antigen by the core 1 β1,3-galactosyltransferase (C1 β3Gal-T or T-synthase) and the resulting Galβ1-3GalNAcα1-Ser/Thr disaccharide corresponds to the T antigen. The C1 β3Gal-T shows a widespread expression in human tissues with predominance in kidney, heart, placenta, and liver [6]. The enzyme activity requires expression of a specific molecular chaperone termed Cosmc (core 1 β3Gal-T-specific molecular chaperone). A somatic mutation in *COSMC* gene has been identified in Tn syndrome leading to a drastic decrease of C1 β3Gal-T activity [7]. Mutations and loss of heterozygosity of *COSMC* gene were also described in colon and melanoma cancer showing Tn-positive cells, as well as in tissues from Tn-positive cervical cancers [8]. However, extensive studies of *COSMC* in epithelial cancers showed that these events were rare and could only partially explain Tn expression in cancers [9]. In normal tissues, the core 1 is a precursor for core 2 *O*-glycans. Three different genes encoding core 2 β1,6-N-acetylglucosaminyltransferases (C2GnT) have been identified in human genome (Table **1**). The main enzyme involved in core 2 biosynthesis is the C2GnT-1 that is expressed in many human tissues [10]. A second C2GnT, named C2/4GnT, is expressed mainly in colon, kidney, pancreas, and small intestine. This enzyme functions in both cores 2 and 4 *O*-glycan branch formation [11]. The third enzyme C2GnT-3 exhibits a unique expression pattern with a high level of expression in thymus but only low levels in other organs, suggesting a specific function different from other members of the C2GnT gene family [12]. An increased expression of C2GnT-has been reported in leukemia [13], lung [14], prostate [15] and pancreas [16] cancers. At the opposite, C2GnT is decreased in colon [17] and breast [18] cancers, the loss of core-2 branching enzyme leading to shorter *O*-glycan chains such as TF antigens. Alternatively, when core 1 is not converted in core 2, it is often the substrate for sialyltransferases adding one or two sialic acid residues masking the T antigen and forming the sialyl-T or disialyl-T. Core 1 can be also extended by the β1,3-N-acetylglucosaminyltransferase 3 (β3GlcNAc-T3), which is highly expressed in the small intestine, colon, and placenta and moderately expressed in various tissues, including the liver, kidney, pancreas, and prostate [19]. The biosynthesis of the core 3 is catalyzed by a unique β1,3-N-acetylglucosaminyltransferase (β3Gn-T6, core 3 synthase), which transfers a GlcNAc residue in β1,3 linkage onto Tn antigen, the expression of β3Gn-T6 being restricted to the stomach, colon, and small intestine [20]. Core 3 synthase is down-regulated in colon carcinoma and profoundly suppresses the metastatic potential of carcinoma cells [21]. As indicated before, the C2/4GnT, functions in both core 2 and core 4 *O*-glycan branch formation [19]. C2/4GnT is frequently down regulated in colorectal cancer and its re-expression causes growth inhibition of colon cancer cells [22].

Sialylation of Tn antigen has been shown to be performed *in vitro* by two members of the sialyltransferase family, namely ST6GalNAc I and ST6GalNAc II [23, 24]. However, studies using cells transfected by either one of these enzymes have demonstrated that in a cellular context, only ST6GalNAc I is able to create STn structures as recognized by anti-STn antibodies [23, 25]. Furthermore, ST6GalNAc I over-expression was shown to correlate with STn expression in gastric, breast and bladder tumors confirming the crucial role of ST6GalNAc I in STn biosynthesis [25 - 27]. It has been proposed that *COSMC* mutation was necessary to provide Tn acceptor substrate for ST6GalNAc I to synthesize STn [8]. Transfection of ST6GalNAc I cDNA has been shown to be sufficient to induce STn expression in various breast cancer cell lines expressing core 1 and core 2 glycans [25, 28, 29], proving that ST6GalNAc I can compete with active T-synthase. Thus, STn expression in cancer is most probably due to over-expression of ST6GalNAc I, with enhancing effects of a decreased competition (*i.e.*, decrease of the expression or activity of the core-synthases).

Expression of Thomsen–Friedenreich Antigens in Cancers

T Antigen

In normal tissues, the T antigen (Galβ1-3GalNAcα1-Ser/Thr) is substituted by sialic acids or by other sugar chains to form more complex *O*-glycans but unsubstituted Galβ1-3GalNAc occurs in about 90% of human cancers including colon, breast, bladder, prostate, liver, ovary and stomach [1]. In many of these cases, the increased T antigen correlates with cancer progression and metastasis [30]. For example, the expression of T antigen is 4 to 6 times higher in invasive compared to non-invasive bladder cancer [31]. In breast cancer, 98% of the disseminated tumor cells in the bone marrow are stained by anti-T antibody, suggesting a role for T antigen in the metastasis process [32].

Tn Antigen

Tn antigen is a pan-carcinoma antigen, expressed on a majority of carcinomas, such as breast, pancreas, colon, lung and bladder, being less common in hematological malignancies. In normal tissues, the Tn is only expressed in embryonic brain [33] and occasionally observed in adult normal cells in the secretory apparatus, but not found on proteins at cell surface or secreted [34, 35]. Over-expression of Tn antigen promotes cancer cell proliferation and invasiveness and patients with Tn-bearing tumors have a worse prognosis for overall and progression-free survival. Tn antigen is also detected at early stages of tumor development and may serve as a biomarker, since its expression is associated with invasive and highly proliferative tumors, and metastasis [36]. Association with high grade has been reported for invasive ductal carcinomas (IDC) using a pool of anti-Tn antibodies [37] and for ductal carcinoma *in situ* using *Vicia villosa* isolectin B4 and *Griffonia simplicifolia* agglutinin [38]. Moreover, it is known that T and Tn antigens are involved in the adhesion of tumor cells to the endothelium *via* a mechanism recruiting Galectin-3 and MUC-1, which is the first step in metastasis formation.

Sialyl-Tn Antigen

Sialyl-Tn is also a pan-carcinoma antigen that is expressed early in tumorigenesis.

In contrast to Tn, STn is not a normal biosynthetic precursor, meaning that its expression is necessarily pathologic [34]. As a side note, in some normal tissues, such as colon, the sialic acid residue of STn may be *O*-acetylated, thus masking the STn and therefore its recognition by anti-STn antibodies [39]. No other sugars are known to be added to the STn antigen. STn is often co-expressed with Tn and therefore mechanisms that result in Tn over-expression likely apply to STn. Specific mechanisms for STn overexpression, include ST6GalNAc-I up-regulation, or re-localization from the Golgi to the endoplasmic reticulum and also loss of *O*-acetyl groups from STn, have been demonstrated in some cancers [26 - 28, 40]. Specifically, STn expression modulates a malignant phenotype inducing a more aggressive cell behavior in gastric and breast carcinoma cells, such as decreased cell-cell aggregation and increased extracellular matrix adhesion, migration and invasion [29, 41 - 43].

GENERAL ASPECTS OF IMMUNE RESPONSE AGAINST CARBOHYDRATE ANTIGENS

The immune system has the greatest potential for the specific destruction of cancer with no toxicity to normal tissue and for long-term memory that can prevent cancer recurrence. However, cancer development is accompanied by a deep immune suppression that affects the effective anti-tumor response and the elimination of the cancer. Evidences have also shown that the aberrant expression of tumor associated carbohydrates is a key factor for cancer immune suppression. Also, the tumor associated carbohydrates tend to be very poorly immunogenic due to their deficient ability to elicit T cell mediated immune responses [44 - 46].

A proper anti-tumor immune response implies two intermeshed main arms: cellular and humoral immune responses.

In cellular immune responses, cytotoxic T cells and Natural killer cells are activated and become able to eliminate tumor cells by apoptosis. For cytotoxic T cell activation, tumor derived peptides have to be presented through the major histocompatibility complex (MHC) class I. One of the downfalls of this mechanism is that MHC presents peptides, so the mechanism is mainly formatted to fight peptide antigens, which can be a limitation in the development of strong immune responses against carbohydrates. It has been reported that cytotoxic T cells may recognize mono- and disaccharides attached to peptides [27, 37 - 40], however further investigation is needed to better understand the relevance of peptide glycosylation in MHC presentation in the context of anti-tumor immune responses.

The humoral immune response involves antibodies specific for tumor cells, which promote a number of effector functions such as antibody dependent cytotoxicity mediated by Natural killer cells, complement dependent cytotoxicity, phagocytosis and antigen neutralization. In contrast to T cells, B cells, through their receptors (*i.e.* immunoglobulins (Ig)) can recognize directly carbohydrate antigens, which will activate the B cells and initiate the secretion of IgM antibodies. IgM is the first immunoglobulin isotype to be secreted by plasma B cells (*i.e.*, effector B cells), characterized by a pentameric structure with relatively low affinity to antigens and short time period, as compared with the other isotypes. This process is called the T cell-independent B cell activation and is able to provide immunity however with limited duration.

In order to have a long lasting anti-tumor humoral response, other antibodies isotypes, besides IgMs are needed, which implies the involvement of helper T (Th) cells in a process called T cell-dependent B cell activation. Antigen specific Th cells are activated by antigen presenting cells, such as dendritic cells (DCs), macrophages and B cells, that uptake tumor antigens and then present derived peptides through MHC class II. Cytokines secreted by Th cells enable the switching of antibody subtypes from IgM to high-affinity IgGs and the differentiation of carbohydrate specific plasma B cells into memory B cells [41]. Then these high-affinity IgG antibodies can bind to the target cancer cells, marking them for destruction by either by Natural killer (NK) cells, complement or macrophages.

In the last few years, a relevant number of reports describing the immunomodulating role of the Tn and STn antigens has elucidated their contribution to immune tolerance. For example, Tn antigen expressed in mucin 6 (MUC6) protein abrogates Th1 cell responses and promotes interleukin 17 (IL-17) response, which might favor immune escape of tumor cells [47]. It was also reported that Tn is recognized by the tolerogenic lectin - macrophage galactose C-type lectin (MGL) - expressed by DCs and macrophages, which endows these cells to suppress T cell immunity, thus playing a role in tumor progression [48]. Cancer mucins can interact with CD22 and down-modulate B cell signal transduction [49]. Moreover, siglec-15 recognizes the STn antigen expressed by cancer cells and transduces a signal for enhanced transforming growth factor beta (TGF-β) secretion by tumor-associated macrophages, which can contribute immunosuppression and to tumor progression by the TGF-β-mediated modulation of intratumoral microenvironment [50]. More recently, it was shown that STn-expressing cancer cells impair DCs' maturation, endowing a tolerogenic function and therefore limiting their capacity to trigger protective anti-tumor T cell responses [51]. In the same study it was observed that blockade of STn antigens expressed by cancer cells was able to lower the induction of tolerance *in vitro* and DCs become more mature. These findings suggest that targeted therapies based on antibodies may provide efficient means to enhance immune responses against STn tumor cells.

These findings lead to the development of new approaches to improve carbohydrate immunogenicity in order to produce more potent cancer immunotherapies. These approaches involved the covalently coupling of carbohydrates to immunologically active protein carriers, such as keyhole limpet hemocyanin (KLH), adjuvants or other immunological epitopes [42].

ANTICANCER VACCINES

The low or absent expression of TF antigens in normal cells turns these structures a potential therapeutic target for anticancer strategies [52, 53]. However, these antigens also participate in the ability of cancer cells to escape the immunological surveillance as the tumor develops, inducing an immunological tolerance to those antigens. Besides the immune tolerance, due to their carbohydrate nature, the antigens are not potent enough to promote a strong immune response against the cancer cells by themselves. Therefore, therapies against them should be robust to overcome the immunological tolerance and restore the immunological surveillance as well as induce long-term memory and protection.

The immunotherapy offers a greater advantage over the usual therapies against cancer, such as chemo- and radiotherapy. Their capacity to boost patient's immune system to fight specifically cancer cells turn the immunotherapy more efficient and reduce adverse effects, compared with other therapies. Importantly, they are able to provide patients with long lasting protection against cancer cells.

In the last decades the role of the immune system and its influence in cancer became better understood, being an important step for the development of therapeutic vaccines against cancer [54]. Furthermore, the advances in several chemistry fields, namely the ones focused on carbohydrates, promoted an intense involvement in the design and development of synthetic carbohydrate-based vaccine candidates, which allowed chemically conjugated constructs to be successfully developed against several kinds of pathogens and widely used in the clinic. The same biotechnological process is being applied in the development of vaccines containing tumor associated carbohydrates and several improvements have been performed aiming that these synthetic vaccine candidates can be clinically relevant [55].

Current vaccines being developed, used in preclinical and clinical settings or being used as therapies are constituted by glycans, either in their native form or in structures derived from it [55]. These vaccines are designed to be tumor-like antigen molecules with the purpose of being used to direct the immune system against cancer cells, promoting their elimination through humoral and/or cell-mediated immunity [54]. Vaccine constructs containing tumor-associated carbohydrate antigens, which are properly exposed to both B and T cells will stimulate the cytotoxic immune responses and production of IgG antibodies against those antigens. This is essential to assist in the destruction of the cancer cells expressing those carbohydrate epitopes. Thus, based on this hypothesis, cancer immunologists and synthetic chemists have been trying to develop anticancer vaccines based on carbohydrate antigens and glycopeptides that are recognized by T cells [56, 57]. However, because carbohydrates antigens are poorly immunogenic, it is necessary to improve their immunogenicity and consequently, the ability of the vaccine candidate to induce a strong immunological response against that target antigen in order to kill the tumor cells. Also, the procedures used for the synthesis or the isolation of the antigens are important for the final vaccine construct [55, 58].

Ideally, cancer vaccines should use antigens that are expressed by cancer cells, and not by normal cells, to guarantee that the immune response that is generated will be targeting only the tumor cells. To improve the antigen's immunogenicity, many times these vaccines are co-administered with adjuvants, which are molecules that can activate cells of the innate immunity to produce cytokines and also activate antigen presenting cells so they can stimulate a strong cellular immune response by the T cells. Adjuvants can have many chemical natures, so they can be emulsions, liposomes and surfactants, for example [59]. Immunogenic carrier molecules, like (KLH) and (BSA) are also frequently used in conjugation with the target antigens. Using this antigen-carrier molecule construction has shown to be the optimal approach for the induction of antibodies against the antigen, when designing these vaccines [52, 53, 58, 60].

Vaccines Against Thomsen–Friedenreich Antigens

Although anticancer vaccines seem a promising approach, to date there is still no approved vaccine for clinical use against TF tumor-associated carbohydrate antigens. This demonstrates the difficulties in discovering the correct target, but also in establishing the most effective immunization, and adjuvant strategies and the most adequate stage of the cancer disease to administer the vaccine and also the right combination with other therapies [61, 62]. However, some attempts have been made in order to design candidate compounds, and some of them were evaluated either in preclinical or in clinical studies (Table **1**) and shown an induction of antigen specific antibodies and the consequent expected immune response [57].

Theratope and Other Monovalent Vaccines

One example of the vaccine candidates mentioned previously is the Theratope®, a vaccine designed by Biomira Inc., a pharmaceutical company now named Oncothyreon from Canada. Theratope® is a therapeutic cancer vaccine consisting of a synthetic STn-mimicking antigen conjugated with KLH and administrated with an adjuvant (Detox B emulsion, later named Enhanzyn™) designed to target metastatic breast cancer. In preclinical studies in mice it was observed that immunization with Theratope® induced STn-specific IgG antibodies and that using a higher conjugation ratio of STn to KLH originated increased specific antibody titers. The formulation of this vaccine was optimized and developed hoping it would increase the immune response in patients against STn breast cancer cells [63 - 66]. In a phase II trial, this vaccine was demonstrated to be safe and immunogenic, generating potent and STn-specific humoral responses that correlated with the overall survival of the patients. In different studies, patients were either treated with cyclophosphamide or submitted to high-dose chemotherapy followed by autologous stem cell rescue, prior to vaccination with Theratope®, with the purpose of overcoming the immune suppressive environment induced by tumor cells. These approaches were well tolerated in the clinical trials [61, 65]. Despite being well tolerated by patients with metastatic breast cancer in a phase III randomized trial (NCT00003638), and also showing a strong and specific humoral response to the STn antigen, no increased overall benefit nor increasing survival was observed in the immunized patients when compared to the control groups, and the vaccine failed to meet the endpoints established. This was probably due to the fact that while the vaccine triggered a B cell-mediated immune response, it failed to trigger a T cell-mediated immune response [63, 66 - 68]. Patients with metastatic breast cancer were administered with the STn-KLH vaccine associated with hormone therapy (NCT00046371). Here it was observed a significant overall survival when compared with the patients receiving hormone therapy only, which suggested that a combination of vaccines and hormone therapy could be favorable [61, 69]. A reinvestigation of the Theratope vaccine trials suggested that the anti-STn antibodies that were produced by patients were probably targeting a variety of STn-carrying glycoproteins, instead of a single one. This raised the question that the identification of glycoproteins bearing STn in cancer cells is a crucial point for our understanding on how the control of tumor growth [70], and the use of relevant STn glycopeptides will be a better formulation for vaccination. Another monoantigenic vaccine targeting prostate cancer (Table **1**) and consisting of the T antigen conjugated to KLH entered a phase I clinical trial (NCT00003819). However, no results are yet posted.

Table 1. Vaccines against TF antigens in clinical trials (as of February 2016).

Brand	Company	Immunogen	Conjugation	Adjuvant	Target (Type of Cancer)	Type of Therapy	Clinical Trials Phase	State	Main Results	References*
Theratope®	Biomira Inc. (now Oncothyreon, Alberta, Canada)	Synthetic STn-mimicking antigen	KLH	Detox B emulsion (later named Enhanzyn)	Metastatic breast cancer	In combination with aromatase inhibitors or Faslodex® treatment	Phase II	Completed	No results posted	NCT00046371
NA		Synthetic STn-mimicking antigen	KLH	Detox B emulsion (later named Enhanzyn)	Metastatic breast cancer	NA	Phase III randomized trial	Completed	No increased overall survival was observed	NCT00003638 [64, 68];
NA	NA	Globo-H-GM2-Lewis-y-MUC1-32(aa)-sTn-TF-Tn	KLH	QS-21	Breast cancer	NA	Not provided	Completed	All 13 participants experienced toxicities, few adverse effects	NCT00030823

(Table 1) contd.....

Brand	Company	Immunogen	Conjugation	Adjuvant	Target (Type of Cancer)	Type of Therapy	Clinical Trials Phase	State	Main Results	References*
NA	NA	Globo-H-GM2-sTn-T-Tn	KLH	QS-21	Epithelial Ovarian, Fallopian Tube, or Peritoneal cancer	NA	Phase I	Active, not yet recruiting participants (Estimated Primary Completion Date: November 2016)	No results yet	NCT01248273
NA	NA	T antigen	KLH	QS-21	Prostate cancer	NA	Phase I	Completed	No results yet	NCT0003819
NA	NA	Globo-H-MUC1-T-Tn	KLH	OPT-821	Epithelial Ovarian, Fallopian Tube, or Primary Peritoneal cancer	In combination with bevacizumab (anti-VEGFA antibody)	Not provided	Active, not recruiting yet participants (Estimated Primary Completion Date: October 2016)	No results yet	NCT01223235

Abbreviations: NA, Not applicable; STn, Sialyl-Thomsen-nouveau; MUC1, Mucin 1; T, Thomsen-Friedenreich; Tn, Thomsen-nouveau; KLH, Keyhole limpet hemocyanin; QS-21 and OPT-821, saponins from the bark of *Quillaja saponaria*.

* Clinical trial number according to Clinicaltrials.gov (https://clinicaltrials.gov/)

Multiantigenic Vaccines

It has been hypothesized by several researchers that a vaccine candidate with multiple tumor antigens could induce a stronger immune response against tumor cells by targeting several antigens at a time. By including the most frequent cancer associated carbohydrates, these multiantigenic vaccines would also overcome the problem of tumor cell heterogeneity.

Thus, several unimolecular multiantigenic vaccines have been synthesized and evaluated in preclinical studies. Ragupathi *et al.* synthesized monovalent peptide constructs attached to immunogenic carrier molecules and then focused on the combination of several carbohydrate antigens associated with a certain type of cancer, due to the heterogeneity of carbohydrate epitopes that cancer cells often express on their surface, with the goal of inducing a stronger immune response against cancer cells. This strategy can be applied by creating a polyvalent monomeric construct, in which mixtures of monomeric-KLH constructs would be injected simultaneously and an antibody response to each antigen would be expected. They demonstrated that an injection of mixtures of monovalent-KLH conjugates as tetravalent (carrying four antigens) or heptavalent (carrying seven antigens) vaccines induced antibody titers for each singular antigen similar to the ones induced by the monovalent vaccines [53].

Kunz *et al.* initiated the synthesis of vaccine constructs containing tumor associated mucin glycopeptides. They synthesized a sialyl-Tn-MUC 1 glycopeptide construct, conjugated with tetanus toxoid (TTox) and immunized mice using complete Freund's adjuvant. The results obtained showed that the construct induced strong and selective immune responses in mice, which was also observed against breast cancer tissues [55, 71].

Recently, Westerlind *et al.* designed and synthesized two vaccine candidates, based on the MUC 1 anticancer vaccines previously designed. The two vaccines consisted of MUC4 tandem repeat glycopeptides containing T or STn antigens, and were conjugated to TTox. The two constructs were evaluated in mice after immunization with complete Freund's adjuvant. High titers of IgG1 antibodies were obtained, indicating that strong responses were generated against the two

MUC4 glycopeptides. The antisera was shown to have its reactivity dependent on the glycosylation site and the antibodies produced showed weak binding to pancreatic tumor cells, probably due to differences in the glycosylation sites [72].

Zhu *et al.* synthesized a unimolecular construct, using five different cancer associated carbohydrate antigens, to target breast and prostate cancers. In the pentavalent construct, the antigens (Globo-H, Ley, STn, TF and Tn) were conjugated to KLH and an adjuvant (QS-21, a naturally occurring saponin isolated from the bark of a tree *Quillaja saponaria* [73]) was used when this construct was evaluated in mice. With exception of Ley, probably because this antigen is endogenously expressed at high levels, the induction of antibodies against the other antigens was observed. These antibodies revealed to be reactive against cancer cell lines expressing the antigens. Therefore, it was observed that even in a pentavalent construct, these antigens maintain their individual immunological properties, which is an advance for the design of carbohydrate-based anticancer vaccines, since several antigens expressed by a cancer cell can be targeted in one vaccine only, increasing its specificity [56, 57].

A new unimolecular pentavalent construct was designed, comprising five carbohydrate antigens overexpressed by breast and prostate cancers - Globo-H, GM2, STn, TF, and Tn. The construct was conjugated to KLH in a high epitope ratio and was also administrated with QS-21 adjuvant. Both IgG and IgM antibodies were produced against the antigens in similar titers using this pentavalent construct than using several monoantigenic vaccines. Also, these antibodies were shown to react with the antigens in breast cancer cell lines [62, 67, 74]. This construct is currently being evaluated in a phase I clinical trial (NCT01248273), in patients with epithelial ovarian, fallopian tube or peritoneal cancer (Table 1).

Structurally Modified Mimetic Vaccines

One reason for the failure of the vaccine candidates in the clinical trials is that the tumor antigens are sensible to endogenous glycosidases, which promotes a decrease in their bioavailability *in vivo*. Several mimetic structures have been therefore proposed in the last years, bearing structural modifications so that they can be more stable against enzymatic degradation than their native structures, without changing their immunogenicity [75]. Besides being resistant to enzymatic degradation, the structural integrity of the saccharide epitope must be maintained in the constructs in order to use tumor associated cancer antigen analogs as vaccines. These analogs must also generate antibodies that present reactivity against the native antigen. Regarding this, one of the strategies used is to fluorinate the compounds, since the C-F binding can mimic effectively the C-OH binding [76].

Richichi *et al.* hypothesized that designing mimetic vaccine candidates instead of using native Tn antigens would promote a higher resistance to enzymatic degradation and therefore a higher bioavailability *in vivo*, which would thus promote stronger immune responses and long-lasting protective efficacy. They designed a new fully synthetic construct, based on previous candidates [75] - a cyclopeptide carrier decorated with clusters of GalNAc, the saccharidic epitope of the Tn antigen, and with either T-helper or chimeric T-helper/T-cell peptide epitopes. Also, this construct contained four residues of a Tn-antigen mimetic and maintained the 4C1 chair conformation of the native antigen. The immunogenicity, protective efficacy and safety of this mucin Tn mimetic antigen were evaluated in mice, after immunization of this construct with CpG1826 adjuvant. It was observed that this vaccine candidate generated a strong and long-lasting humoral response and protection. The antibodies produced were shown to recognize and bind to tumor cells that express the native antigen on their surface, showing a relevant specificity. Furthermore, immunization with this synthetic candidate promoted the reduction of tumor diameter and a higher survival in mice, evidencing the immunotherapeutic effect of this compound [77].

Another synthetic construct, appropriate for application in humans, using the MUC1 glycopeptide and containing a STn side chain, conjugated with the tetanus toxoid protein caused a strong immune response. Similarly, mice that were treated with a similar synthetic construct containing a T-antigen instead of STn showed a selective immune response [78]. Based on this, Hoffmann-Röder *et al.* synthesized a vaccine candidate that used the MUC1 glycopeptide containing a T-antigen side chain, conjugated with the tetanus toxoid (TTox) protein (structure named T-antigen-MUC1-TTox). Moreover, an analogue of the T-antigen carrying fluorine substituents in the 6- and 6'- positions of the saccharide was built, creating then a mimicking T-antigen structure (named F2T-antigen-MUC1-TTox). These structures were also designed to evaluate if the vaccines based on carbohydrates sensitive to enzymatic degradation can be replaced by mimicking structures that are more resistant to degradation. To verify the immunological properties of these two constructs, mice were immunized with them, using complete Freund's adjuvant [18]. It was observed that a strong immune response occurred for both constructs. The natural and fluorinated constructs were also conjugated with (BSA) protein (structures named T-antigen-MUC1-BSA and F2T-antigen-MUC1-BSA). In both cases, either in conjugation with TTox or with BSA, the binding of antibodies to the natural T-antigen-MUC1-BSA/TTox conjugate was similar to the binding of antibodies to the fluoro-substituted conjugate. The antibodies produced were mostly of the IgG1 isotype and almost no IgM antibodies were produced, which indicated that the immune responses were selective and promoted the stimulation of immunological memory. Moreover, it was observed that the IgG antibodies that were generated cross-reacted with native TF epitopes on breast cancer cells. It was concluded that if the OH groups of the T-antigen carbohydrate are replaced by fluorine, a vaccine candidate more resistant to enzymatic degradation can be used, because its immunogenicity was not reduced and the fluorinated construct is still capable of inducing a selective immune response [62, 76, 78, 79].

Furthermore, Yang *et al.* reported that fluorinated STn antigen conjugates showed higher immunogenicity than their native compounds and the antibodies produced cross-reacted with STn positive tumor cells, as expected. Based on these results, Hoffmann-Röder *et al.* reported a synthesis of two new MUC1 glycoconjugate analogs, including a 4'-deoxy-4'-fluoro-TF antigen. These analogs were conjugated with tetanus toxoid (TTox) and (BSA). Besides having an improved immunogenicity, as mentioned before, these fluorinated tumor antigens may also be less prone for enzymatic degradation, increasing the compound's bioavailability. Taking that into account, the hydrolytic resistance of these fluorinated compounds was evaluated and the results showed an increased resistance to enzymatic degradation when compared to the native TF derivative. The immunological properties were evaluated in mice after immunization with

Freund's adjuvant and there was an induction of IgG antibodies that were shown to cross-react with the native antigen on tumor cells [76, 79].

Self-Adjuvanting Vaccines

Although synthetic conjugated constructs have potential to be used in the future as therapeutic anticancer vaccines, it is sometimes known that the immunogenicity of the carrier protein itself can generate a suppression of the immune response to the target antigen. Therefore, self-adjuvanting vaccines have been proposed [62]. Renaudet *et al.* described the first example of a self-adjuvanting vaccine, composed of a cluster of Tn B-cell epitope, a T-helper cell peptide epitope, a T-cytotoxic cell peptide epitope and palmitic acid as a toll-like receptor 2 (TLR2) agonist. Studies on immunized mice without any other external adjuvant showed that this construct promoted a strong immune response that succeeded in an increased survival with tumor regression [62, 80]. Boons *et al.* reported the synthesis of a vaccine candidate, composed of glycopeptide 1 (a MUC1 glycopeptide that contains STn), a T-helper cell epitope form polio virus and the Pam3CysSK4 TLR2 ligand, since this ligand can enhance inflammation locally, promoting the activation of adaptive immunity. They demonstrated that this three-component synthetic self-adjuvanting vaccine candidate generated strong humoral (high IgG antibody titers) and cellular immune responses in mammary cancer mouse model [55, 62, 81].

To conclude, the development of anticancer vaccines still have a long way to improve and goals to achieve. There are still some obstacles to overcome. All the economic processes in designing and testing these vaccine candidates, as well as the growing knowledge about the functioning of the immune system and about the cancers themselves are key points for the development of these potentially effective therapies. Newly designed compounds must also be studied in parallel with other currently used therapies and many aspects still have to be studied and improved to avoid the induction of immunological tolerance or, in contrast, serious side effects and to ultimately improve the overall survival of the patients [54].

ANTIBODY BASED THERAPY AGAINST TF-ANTIGENS

Antibody-based treatments have been widely used as therapeutic strategies due to their exclusive features, such as high specificity and engagement with the immune system [82, 83]. Innovative next generation antibodies with higher efficacy, safety and broader applications have been heightened by the continuous development and optimization of methods involved in antibodies production and engineering, squired by the increasing knowledge on the crucial interplay between cancer cells, antibodies and immune system [84 - 86]. Currently, around 47 products derived from mAbs have been approved in the United States and Europe for the treatment of a wide variety of diseases, ranging from cancer to infectious and cardiovascular diseases to autoimmune diseases resulting from a significant growth of therapeutic antibodies within the healthcare industry [87, 88].

Particularly in the cancer field, therapeutic antibodies have been widely and successfully used, representing one of the most important strategies to treat patients with this disease. In their early development, issues associated with immunogenicity or allergic-like reactions were observed for the treatment using mouse derived antibodies. As a result, a next generation of therapeutic antibodies with a more favorable safety profile and half-life raised, based on genetic engineering of mouse variable regions with human constant domains retaining the specificity of the initial developed murine antibodies [89].

Therapeutic Antibodies in Cancer

There are several mechanisms of action associated to antibodies besides the characteristic affinity and specificity to target antigens. In general, antibodies can mediate cell death through the activation of the complement system (complement dependent cytotoxicity (CDC)), redirection of immune effector cells (antibody-dependent cell-mediated cytotoxicity (ADCC)) and by exerting a blocking action on soluble mediators and specific molecules. Both CDC and ADCC are mechanisms that rely on the engagement of the Fc region of antibodies to the complement system or to the Fc receptors expressed on immune effector cells to promote target cell death, respectively [90 - 92]. Additionally, antibodies can induce crosslinking of receptors that are associated to mediators of cell apoptosis, like caspases, which will lead to cell death [93]. These intrinsic antibodies features are important mechanisms of action and have been applied therapeutically in many different fields.

More recently, therapeutic antibodies have been molecularly engineered to adjust antibodies effector functions, size and formats, multimerization and immunogenicity contributing to the development of new and more effective antibody-based therapies [94]. A promising and rapidly growing field is antibody-drug conjugates (ADC) in which the antibody is used as a delivery vehicle for highly potent cytotoxic molecules with specificity for tumor associated antigens [95, 96]. In order to circumvent the limitations of conventional monospecific therapies and achieve enhanced therapeutic efficacy, advances in antibody engineering technologies enabled the development of approaches that target multiple receptors simultaneously. In recent years, a range of multi target antigen-binding and immune cell recruiting therapeutic antibodies have emerged holding great therapeutic potential [97].

In particular, bispecific antibodies (BsAb) in cancer therapy are emerging as a novel approach and two BsAb were already approved for therapy and more than 30 are in clinical trials, most of them based on the retarget of T cells to eradicate tumor cells. Furthermore, various formats of BsAb have emerged and are undergoing clinical evaluation using either IgG fragments or whole IgG molecules, ranging from Bi-specific T-cell engager (BiTEs), Dual-Affinity Re-Targeting (DART) and trifunctional BsAbs [98].

Therapeutic Antibodies Against Tn and STn Antigens

Several antibodies have already been produced against TFs antigens, however those have not proven to be efficient in activating the immune system and eliminating the malignant cells. Targeting the tumor glycan signature thus provides an attractive strategy for immunotherapy and encouragement has been given to promote the development of more effective antibodies against these type of antigens to specifically kill cancer cells [45].

Development of anti-carbohydrate antibodies is considered a demanding task since a T-cell-independent response to carbohydrates is commonly observed, which leads to the production of low affinity IgM antibodies and further complications with screening technologies [99]. Nevertheless, since the early 80´s several antibodies have been developed against glycan antigens such as T, Tn and STn (Table **2**).

T Antigen and Anti-T mAbs

Several mAbs to the T-antigen have been developed and some of them can be found on Table **2**. Among these antibodies it is important to highlight the development of antibodies with different isotypes and species ranging from IgM, IgG and with particular interest single chain variable fragments (scFv). The recent last-mentioned antibody has been proved to specifically recognize T-antigen disaccharide units in various T-antigen presenting conjugates [121]. Moreover, using the same antigen it was possible to develop two antibodies with different isotypes and different specificities against same antigens [103, 122].

Table 2. Monoclonal antibodies against Thomsen–Friedenreich antigens.

Tumor Antigen	mAb Name	mAb Species and Isotype	Immunogen	Reference
T Antigen	**JAA-F11** (ATCC® CRL-2381™)	Mouse IgG3	Synthetic T antigen -BSA conjugate	[100, 101]
	A78-G/ A7	Mouse IgM	Human asialoglycophorin (essentially A)	[102]
	HH8	Mouse IgM	Galactosyl-A glycolipids	[103]
	3C9	Mouse IgG	Galactosyl-A glycolipids	[103]
	1E8	Single chain variable fragment (scFv)	Neoglycolipid T-antigen E6-BDB (probe for a phage library displaying human single chain antibodies)	[104]
	TF1	Human IgM	*In vitro* asialoglycophorin (AGP)-stimulated Leu-Leu-OMe-treated lymphocytes	[105]
	TF2	Human IgA	*In vitro* asialoglycophorin (AGP)-stimulated Leu-Leu-OMe-treated lymphocytes	[105]

(Table 2) contd.....

Tumor Antigen	mAb Name	mAb Species and Isotype	Immunogen	Reference
Tn Antigen	2D9	Mouse IgG1	15Tn-MUC1 60-mer glycopeptide conjugated to KLH	[106]
	5E5	Mouse IgG1	Complete Tn glycosylated MUC1 glycopeptide	[107]
	5F4	Mouse IgM	Enzymatically desialylated ovine submaxillary mucin (aOSM)	[108]
	83D4	Mouse IgM	Cell suspensions from sections of formol-fixed paraffin embedded human breast cancer	[109]
	HBTn1/ HB-Tn1	Mouse IgM (κ)	aOSM	[110]
	MLS128	Mouse IgG3 (κ)	Human colonic cancer cells LS 180	[111]
	KM3413	Mouse IgG1	Mucins purified from a culture supernatant of the human colon cancer cell line LS180	[112]
	BRIC111	Mouse IgG1	Tn red blood cells	[113]
	BRIC66	Mouse IgM	Ovarian cyst blood group A_1 glycoprotein	[113]
	2154F12A4	Mouse IgM	Tn clustered on cationized BSA	[114]
	GOD3-2C4	IgG1 (κ)	A549 cells and boost with the Tn carrying OSM protein	[115]
STn Antigen	HB-STn/ 3F1	Mouse IgG1	Ovine submaxillary mucin (OSM)	[Clausen, unpublished]
	B72.3	Mouse IgG1	Membrane-enriched fraction of a human breast carcinoma liver metastasis	[116]
	LLU9B4	Mouse IgG	STn antigen isolated from human colon adenocarcinoma cell line LS174T grown in nude rats	[117]
	3P9	Mouse IgM (κ)	Human colorectal adenocarcinoma SW1116 cells	[118]
	CC49	Mouse IgG1	mAb B72.3 reactive tumor-associated glycoprotein 72 (TAG-72), isolated from human colon cancer cells	[119]
	MLS132/ MLS102	Mouse IgG3	Human cancer cells (LS 180 cells)	[120]
	TKH2	Mouse IgG1	OSM	[103]

Denoted with therapeutic potential is the highly specific anti-TF antibody JAA-F11 which is currently undergoing preclinical trials for human breast cancer. This antibody has been proved to inhibit human and mouse tumor cell growth *in vitro*; to inhibit metastasis formation in several models of human breast, colon and prostate cancer cells; selectively bind to both human and mouse tumor cells *in vivo* and internalize within 1h in tumor cells. These features highlight the potential of a humanized JAA-F11 antibody to be able to function through direct killing as well as blocking tumor metastasis. Particularly, since this antibody has been shown to internalize rapidly, it may also be an antibody–drug conjugate for cancer therapy [101].

Tn Antigen and Anti-Tn mAbs

Similarly to anti-TF antibodies, several mAbs with different specificities towards the Tn antigen have also been generated using different immunogens (Table **2**). Particular importance should be given to monoclonal antibodies MLS128, KM3413, GOD3-2C4 and 2154F12A4 which have been shown to have promising therapeutic potential against Tn-positive cancers. MLS128 has the ability to inhibit the growth of certain cancer cell lines [123]; the development of a mouse-human chimeric IgG1 of KM3413 (cKM3413) revealed that this chimeric antibody could induce ADCC and direct killing activity against Jurkat cells [112]; GOD3-2C4 also presents ADCC activity against Jurkat cells *in vitro* particularly limiting the growth of xenografts of a human lung carcinoma in SCID mice [115] and 2154F12A4 antibody was able to inhibit cancer cell adhesion to lymphatic endothelium suggesting a novel involvement of Tn in the lymphatic dissemination of cancer cells suggesting future applications in inhibiting specifically lymphatic metastases [114].

STn Antigen and Anti-STn mAbs

The first anti-STn specific mAb, B72.3 [116], was developed in 1981 and since then more antibodies against the glycan STn antigen have been produced. Antibodies with different affinities and specificities towards this antigen were produced using a great diversity of antigens. For instance, the second generation antibody CC49 was obtained by immunization of mice with TAG-72 specific molecules obtained from purification using B72.3 mAb [119]. Furthermore, this antibody and its humanized format are currently undergoing a clinical trial for its use in radioimmunoguided surgery. However, the fine specificity of many mAbs is unclear and STn configuration may be highly important for antibody recognition. For instance, B72.3 strongly binds glycoproteins bearing STn-trimers, but poorly interacts with monomeric-STn glycoproteins [124] and MLS20 was reported to be specific for clustered STn [120]. LLU9B4 and 3P9 are recently developed anti-STn mAbs generated by generated by immunization of mice with STn-positive cell line [117, 118]. Particular interest has been given to the last mentioned antibody since it is an IgM mAb that showed significant inhibition on proliferation and migration of STn expressing cells and tumor growth suggesting its potential application in antibody-based tumor therapy [118].

Future Perspective

In spite of the many advantages in targeting glycans in cancer cells and recent advances in cancer cell glycomics, the number of cancer glycan-specific antibodies with clinical potential is limited. Glycan antibodies specificity is still debatable and the potential therapeutic effect on cancer cells and patient tissues needs to be thoroughly assessed prior to use in therapeutic approaches [45]. In addition, the majority of anti-glycan antibodies are carbohydrate reactive, which means that they react with any glycoprotein carrying the glycan and only few were able to recognize glycan/peptide epitopes [125] though with low affinity. Additionally, development of novel glycan immunization strategies could improve specificity and avoid binding to irrelevant extracellular glycoproteins, which may deceive effective binding of antibodies to cell surface. Antibodies with therapeutic potential against glycans are an emerging field, particularly glycan-based strategies for cargo delivery and immunomodulation have been considered a promising new approach.

CONCLUSION

The TF antigens are potential targets for immunotherapy due to their limited expression in cancer cells. A number of studies have shed light into the mechanisms by which TF-expressing cancer cells show a higher ability to evade immune responses, which helped to foresee more efficient strategies to overcome them and improve treatment. Vaccines have been constructed based on TF structure or as multiantigens combining TF antigens with other cancer associated antigens, aiming to treat different types of cancers. These are already being used in clinical trials and several other that are still being studied at preclinical settings are soon expected to reach clinical stages. Alternatively, antibodies which have proved to be efficient to discriminate between malignant and normal adult cells are now being reinvented for therapeutic applications. Thus TF-based therapeutic antibodies are expected to be used in therapy very soon. A combination of these two approaches, vaccines and antibodies against TF antigens, and also a combination with other immunotherapeutics seems a near future with encouraging perspectives for cancer patients.

CONFLICT OF INTEREST

The authors confirm that they have no conflict of interest to declare for this publication.

ACKNOWLEDGEMENTS

Declared none.

REFERENCES

[1]   Yu LG. The oncofetal Thomsen-Friedenreich carbohydrate antigen in cancer progression. Glycoconj J 2007; 24(8): 411-20.
[http://dx.doi.org/10.1007/s10719-007-9034-3] [PMID: 17457671]

[2]   Springer GF. T and Tn, general carcinoma autoantigens. Science 1984; 224(4654): 1198-206.
[http://dx.doi.org/10.1126/science.6729450] [PMID: 6729450]

[3]   Berger EG. Tn-syndrome. Biochim Biophys Acta 1999; 1455(2-3): 255-68.
[http://dx.doi.org/10.1016/S0925-4439(99)00069-1] [PMID: 10571017]

[4]   Julien S, Delannoy P. Sialyl-Tn antigen in cancer: from diagnosis to therapy. Rec Res Dev Cancer 2003; pp. 185-99.

[5]   Ten Hagen KG, Fritz TA, Tabak LA. All in the family: the UDP-GalNAc:polypeptide N-acetylgalactosaminyltransferases. Glycobiology 2003; 13(1): 1R-16R.
[http://dx.doi.org/10.1093/glycob/cwg007] [PMID: 12634319]

[6]   Ju T, Brewer K, D'Souza A, Cummings RD, Canfield WM. Cloning and expression of human core 1 β1,3-galactosyltransferase. J Biol Chem 2002; 277(1): 178-86.
[http://dx.doi.org/10.1074/jbc.M109060200] [PMID: 11677243]

[7]   Ju T, Cummings RD. Protein glycosylation: chaperone mutation in Tn syndrome. Nature 2005; 437(7063): 1252.

[http://dx.doi.org/10.1038/4371252a] [PMID: 16251947]

[8]   Ju T, Lanneau GS, Gautam T, *et al.* Human tumor antigens Tn and sialyl Tn arise from mutations in Cosmc. Cancer Res 2008; 68(6): 1636-46.
       [http://dx.doi.org/10.1158/0008-5472.CAN-07-2345] [PMID: 18339842]

[9]   Yoo NJ, Kim MS, Lee SH. Absence of COSMC gene mutations in breast and colorectal carcinomas. APMIS 2008; 116(2): 154-5.
       [http://dx.doi.org/10.1111/j.1600-0463.2008.00965.x] [PMID: 18321367]

[10]   Bierhuizen MF, Fukuda M. Expression cloning of a cDNA encoding UDP-GlcNAc:Gal beta 1-3-Gal-NAc-R (GlcNAc to GalNAc) beta 1-6GlcNAc transferase by gene transfer into CHO cells expressing polyoma large tumor antigen. Proc Natl Acad Sci USA 1992; 89(19): 9326-30.
       [http://dx.doi.org/10.1073/pnas.89.19.9326] [PMID: 1329093]

[11]   Schwientek T, Nomoto M, Levery SB, *et al.* Control of O-glycan branch formation. Molecular cloning of human cDNA encoding a novel beta1,6-N-acetylglucosaminyltransferase forming core 2 and core 4. J Biol Chem 1999; 274(8): 4504-12.
       [http://dx.doi.org/10.1074/jbc.274.8.4504] [PMID: 9988682]

[12]   Schwientek T, Yeh JC, Levery SB, *et al.* Control of O-glycan branch formation. Molecular cloning and characterization of a novel thymus-associated core 2 beta1, 6-n-acetylglucosaminyltransferase. J Biol Chem 2000; 275(15): 11106-13.
       [http://dx.doi.org/10.1074/jbc.275.15.11106] [PMID: 10753916]

[13]   Brockhausen I, Kuhns W, Schachter H, Matta KL, Sutherland DR, Baker MA. Biosynthesis of O-glycans in leukocytes from normal donors and from patients with leukemia: increase in O-glycan core 2 UDP-GlcNAc:Gal β 3 GalNAc α-R (GlcNAc to GalNAc) β(1-6)-N-acetylglucosaminyltransferase in leukemic cells. Cancer Res 1991; 51(4): 1257-63.
       [PMID: 1997166]

[14]   Machida E, Nakayama J, Amano J, Fukuda M. Clinicopathological significance of core 2 beta1,6-N-acetylglucosaminyltransferase messenger RNA expressed in the pulmonary adenocarcinoma determined by *in situ* hybridization. Cancer Res 2001; 61(5): 2226-31.
       [PMID: 11280791]

[15]   Hagisawa S, Ohyama C, Takahashi T, *et al.* Expression of core 2 β1,6-N-acetylglucosaminyltransfer-ase facilitates prostate cancer progression. Glycobiology 2005; 15(10): 1016-24.
       [http://dx.doi.org/10.1093/glycob/cwi086] [PMID: 15932919]

[16]   Beum PV, Singh J, Burdick M, Hollingsworth MA, Cheng PW. Expression of core 2 beta-1,6-N-acetylglucosaminyltransferase in a human pancreatic cancer cell line results in altered expression of MUC1 tumor-associated epitopes. J Biol Chem 1999; 274(35): 24641-8.
       [http://dx.doi.org/10.1074/jbc.274.35.24641] [PMID: 10455130]

[17]   Yang J-M, Byrd JC, Siddiki BB, *et al.* Alterations of O-glycan biosynthesis in human colon cancer tissues. Glycobiology 1994; 4(6): 873-84.
       [http://dx.doi.org/10.1093/glycob/4.6.873] [PMID: 7734850]

[18]   Dalziel M, Whitehouse C, McFarlane I, *et al.* The relative activities of the C2GnT1 and ST3Gal-I glycosyltransferases determine O-glycan structure and expression of a tumor-associated epitope on MUC1. J Biol Chem 2001; 276(14): 11007-15.

[http://dx.doi.org/10.1074/jbc.M006523200] [PMID: 11118434]

[19]   Yeh J-C, Hiraoka N, Petryniak B, *et al.* Novel sulfated lymphocyte homing receptors and their control by a Core1 extension β 1,3-N-acetylglucosaminyltransferase. Cell 2001; 105(7): 957-69.
[http://dx.doi.org/10.1016/S0092-8674(01)00394-4] [PMID: 11439191]

[20]   Iwai T, Inaba N, Naundorf A, *et al.* Molecular cloning and characterization of a novel UDP-GlcNAc:GalNAc-peptide beta1,3-N-acetylglucosaminyltransferase (beta 3Gn-T6), an enzyme synthesizing the core 3 structure of O-glycans. J Biol Chem 2002; 277(15): 12802-9.
[http://dx.doi.org/10.1074/jbc.M112457200] [PMID: 11821425]

[21]   Iwai T, Kudo T, Kawamoto R, *et al.* Core 3 synthase is down-regulated in colon carcinoma and profoundly suppresses the metastatic potential of carcinoma cells. Proc Natl Acad Sci USA 2005; 102(12): 4572-7.
[http://dx.doi.org/10.1073/pnas.0407983102] [PMID: 15755813]

[22]   Huang MC, Chen HY, Huang HC, *et al.* C2GnT-M is downregulated in colorectal cancer and its re-expression causes growth inhibition of colon cancer cells. Oncogene 2006; 25(23): 3267-76.
[http://dx.doi.org/10.1038/sj.onc.1209350] [PMID: 16418723]

[23]   Marcos NT, Pinho S, Grandela C, *et al.* Role of the human ST6GalNAc-I and ST6GalNAc-II in the synthesis of the cancer-associated sialyl-Tn antigen. Cancer Res 2004; 64(19): 7050-7.
[http://dx.doi.org/10.1158/0008-5472.CAN-04-1921] [PMID: 15466199]

[24]   Harduin-Lepers A, Vallejo-Ruiz V, Krzewinski-Recchi M-A, Samyn-Petit B, Julien S, Delannoy P. The human sialyltransferase family. Biochimie 2001; 83(8): 727-37.
[http://dx.doi.org/10.1016/S0300-9084(01)01301-3] [PMID: 11530204]

[25]   Sewell R, Bäckström M, Dalziel M, *et al.* The ST6GalNAc-I sialyltransferase localizes throughout the Golgi and is responsible for the synthesis of the tumor-associated sialyl-Tn O-glycan in human breast cancer. J Biol Chem 2006; 281(6): 3586-94.
[http://dx.doi.org/10.1074/jbc.M511826200] [PMID: 16319059]

[26]   Marcos NT, Bennett EP, Gomes J, *et al.* ST6GalNAc-I controls expression of sialyl-Tn antigen in gastrointestinal tissues. Front Biosci (Elite Ed) 2011; 3: 1443-55.
[PMID: 21622148]

[27]   Ferreira JA, Videira PA, Lima L, *et al.* Overexpression of tumour-associated carbohydrate antigen sialyl-Tn in advanced bladder tumours. Mol Oncol 2013; 7(3): 719-31.
[http://dx.doi.org/10.1016/j.molonc.2013.03.001] [PMID: 23567325]

[28]   Julien S, Krzewinski-Recchi M-A, Harduin-Lepers A, *et al.* Expression of sialyl-Tn antigen in breast cancer cells transfected with the human CMP-Neu5Ac: GalNAc α2,6-sialyltransferase (ST6GalNac I) cDNA. Glycoconj J 2001; 18(11-12): 883-93.
[http://dx.doi.org/10.1023/A:1022200525695] [PMID: 12820722]

[29]   Julien S, Lagadec C, Krzewinski-Recchi MA, Courtand G, Le Bourhis X, Delannoy P. Stable expression of sialyl-Tn antigen in T47-D cells induces a decrease of cell adhesion and an increase of cell migration. Breast Cancer Res Treat 2005; 90(1): 77-84.
[http://dx.doi.org/10.1007/s10549-004-3137-3] [PMID: 15770530]

[30]   Wolf MF, Ludwig A, Fritz P, Schumacher K. Increased expression of Thomsen-Friedenreich antigens

during tumor progression in breast cancer patients. Tumour Biol 1988; 9(4): 190-4.
[http://dx.doi.org/10.1159/000217561] [PMID: 3420374]

[31] Langkilde NC, Wolf H, Clausen H, Kjeldsen T, Ørntoft TF. Nuclear volume and expression of T-antigen, sialosyl-Tn-antigen, and Tn-antigen in carcinoma of the human bladder. Relation to tumor recurrence and progression. Cancer 1992; 69(1): 219-27.
[http://dx.doi.org/10.1002/1097-0142(19920101)69:1<219::AID-CNCR2820690136>3.0.CO;2-A]
[PMID: 1727666]

[32] Schindlbeck C, Jeschke U, Schulze S, *et al.* Characterisation of disseminated tumor cells in the bone marrow of breast cancer patients by the Thomsen-Friedenreich tumor antigen. Histochem Cell Biol 2005; 123(6): 631-7.
[http://dx.doi.org/10.1007/s00418-005-0781-6] [PMID: 15889266]

[33] Akita K, Fushiki S, Fujimoto T, *et al.* Developmental expression of a unique carbohydrate antigen, Tn antigen, in mouse central nervous tissues. J Neurosci Res 2001; 65(6): 595-603.
[http://dx.doi.org/10.1002/jnr.1190] [PMID: 11550228]

[34] Kudelka MR, Ju T, Heimburg-Molinaro J, Cummings RD. Simple sugars to complex disease--muc-n-type O-glycans in cancer. Adv Cancer Res 2015; 126: 53-135.
[http://dx.doi.org/10.1016/bs.acr.2014.11.002] [PMID: 25727146]

[35] Osako M, Yonezawa S, Siddiki B, *et al.* Immunohistochemical study of mucin carbohydrates and core proteins in human pancreatic tumors. Cancer 1993; 71(7): 2191-9.
[http://dx.doi.org/10.1002/1097-0142(19930401)71:7<2191::AID-CNCR2820710705>3.0.CO;2-X]
[PMID: 8384065]

[36] Itzkowitz SH, Bloom EJ, Lau TS, Kim YS. Mucin associated Tn and sialosyl-Tn antigen expression in colorectal polyps. Gut 1992; 33(4): 518-23.
[http://dx.doi.org/10.1136/gut.33.4.518] [PMID: 1582597]

[37] Wang B-L, Springer GF, Carlstedt SC. Quantitative computerized image analysis of Tn and T (Thomsen-Friedenreich) epitopes in prognostication of human breast carcinoma. J Histochem Cytochem 1997; 45(10): 1393-400.
[http://dx.doi.org/10.1177/002215549704501007] [PMID: 9313800]

[38] Korourian S, Siegel E, Kieber-Emmons T, Monzavi-Karbassi B. Expression analysis of carbohydrate antigens in ductal carcinoma *in situ* of the breast by lectin histochemistry. BMC Cancer 2008; 8(1): 136.
[http://dx.doi.org/10.1186/1471-2407-8-136] [PMID: 18479514]

[39] Itzkowitz SH, Yuan M, Montgomery CK, *et al.* Expression of Tn, sialosyl-Tn, and T antigens in human colon cancer. Cancer Res 1989; 49(1): 197-204.
[PMID: 2908846]

[40] Vázquez-Martín C, Cuevas E, Gil-Martín E, Fernández-Briera A. Correlation analysis between tumor-associated antigen sialyl-Tn expression and ST6GalNAc I activity in human colon adenocarcinoma. Oncology 2004; 67(2): 159-65.
[http://dx.doi.org/10.1159/000081003] [PMID: 15539921]

[41] Pinho S, Marcos NT, Ferreira B, *et al.* Biological significance of cancer-associated sialyl-Tn antigen: modulation of malignant phenotype in gastric carcinoma cells. Cancer Lett 2007; 249(2): 157-70.

[http://dx.doi.org/10.1016/j.canlet.2006.08.010] [PMID: 16965854]

[42]   Julien S, Adriaenssens E, Ottenberg K, *et al.* ST6GalNAc I expression in MDA-MB-231 breast cancer cells greatly modifies their O-glycosylation pattern and enhances their tumourigenicity. Glycobiology 2006; 16(1): 54-64.
[http://dx.doi.org/10.1093/glycob/cwj033] [PMID: 16135558]

[43]   Tamura F, Sato Y, Hirakawa M, *et al.* RNAi-mediated gene silencing of ST6GalNAc I suppresses the metastatic potential in gastric cancer cells. Gastric Cancer 2016; 19(1): 85-97.
[http://dx.doi.org/10.1007/s10120-014-0454-z] [PMID: 25532910]

[44]   Julien S, Videira PA, Delannoy P. Sialyl-tn in cancer: (how) did we miss the target? Biomolecules 2012; 2(4): 435-66.
[http://dx.doi.org/10.3390/biom2040435] [PMID: 24970145]

[45]   Loureiro LR, Carrascal MA, Barbas A, *et al.* Challenges in antibody development against Tn and Sialyl-Tn antigens. Biomolecules 2015; 5(3): 1783-809.
[http://dx.doi.org/10.3390/biom5031783] [PMID: 26270678]

[46]   Monzavi-Karbassi B, Pashov A, Kieber-Emmons T. Tumor-associated glycans and immune surveillance. Vaccines (Basel) 2013; 1(2): 174-203.
[http://dx.doi.org/10.3390/vaccines1020174] [PMID: 26343966]

[47]   Freire T, Lo-Man R, Bay S, Leclerc C. Tn glycosylation of the MUC6 protein modulates its immunogenicity and promotes the induction of Th17-biased T cell responses. J Biol Chem 2011; 286(10): 7797-811.
[http://dx.doi.org/10.1074/jbc.M110.209742] [PMID: 21193402]

[48]   Saeland E, van Vliet SJ, Bäckström M, *et al.* The C-type lectin MGL expressed by dendritic cells detects glycan changes on MUC1 in colon carcinoma. Cancer Immunol Immunother 2007; 56(8): 1225-36.
[http://dx.doi.org/10.1007/s00262-006-0274-z] [PMID: 17195076]

[49]   Toda M, Akita K, Inoue M, Taketani S, Nakada H. Down-modulation of B cell signal transduction by ligation of mucins to CD22. Biochem Biophys Res Commun 2008; 372(1): 45-50.
[http://dx.doi.org/10.1016/j.bbrc.2008.04.175] [PMID: 18474217]

[50]   Takamiya R, Ohtsubo K, Takamatsu S, Taniguchi N, Angata T. The interaction between Siglec-15 and tumor-associated sialyl-Tn antigen enhances TGF-β secretion from monocytes/macrophages through the DAP12-Syk pathway. Glycobiology 2013; 23(2): 178-87.
[http://dx.doi.org/10.1093/glycob/owe139] [PMID: 23035012]

[51]   Carrascal MA, Severino PF, Guadalupe Cabral M, *et al.* Sialyl Tn-expressing bladder cancer cells induce a tolerogenic phenotype in innate and adaptive immune cells. Mol Oncol 2014; 8(3): 753-65.
[http://dx.doi.org/10.1016/j.molonc.2014.02.008] [PMID: 24656965]

[52]   Keding SJ, Danishefsky SJ. Prospects for total synthesis: a vision for a totally synthetic vaccine targeting epithelial tumors. Proc Natl Acad Sci USA 2004; 101(33): 11937-42.
[http://dx.doi.org/10.1073/pnas.0401894101] [PMID: 15280546]

[53]   Ragupathi G, Koide F, Livingston PO, *et al.* Preparation and evaluation of unimolecular pentavalent and hexavalent antigenic constructs targeting prostate and breast cancer: a synthetic route to anticancer

vaccine candidates. J Am Chem Soc 2006; 128(8): 2715-25.
[http://dx.doi.org/10.1021/ja057244+] [PMID: 16492059]

[54]   Liu JK. Anti-cancer vaccines - a one-hit wonder? Yale J Biol Med 2014; 87(4): 481-9.
[PMID: 25506282]

[55]   Fernández-Tejada A, Cañada FJ, Jiménez-Barbero J. Recent Developments in Synthetic
Carbohydrate-Based Diagnostics, Vaccines, and Therapeutics. Chemistry 2015; 21(30): 10616-28.
[http://dx.doi.org/10.1002/chem.201500831] [PMID: 26095198]

[56]   Zhu J, Wan Q, Lee D, *et al.* From synthesis to biologics: preclinical data on a chemistry derived
anticancer vaccine. J Am Chem Soc 2009; 131(26): 9298-303.
[http://dx.doi.org/10.1021/ja901415s] [PMID: 19518111]

[57]   Lee D, Danishefsky SJ. "Biologic" level structures through chemistry: A total synthesis of a
unimolecular pentavalent MUCI glycopeptide construct. Tetrahedron Lett 2009; 50(19): 2167-70.
[http://dx.doi.org/10.1016/j.tetlet.2009.02.138] [PMID: 21423786]

[58]   Pejawar-Gaddy S, Finn OJ. Cancer vaccines: accomplishments and challenges. Crit Rev Oncol
Hematol 2008; 67(2): 93-102.
[http://dx.doi.org/10.1016/j.critrevonc.2008.02.010] [PMID: 18400507]

[59]   Banday AH, Jeelani S, Hruby VJ. Cancer vaccine adjuvants--recent clinical progress and future
perspectives. Immunopharmacol Immunotoxicol 2015; 37(1): 1-11.
[http://dx.doi.org/10.3109/08923973.2014.971963] [PMID: 25318595]

[60]   Huang Z-H, Sun Z-Y, Gao Y, *et al.* Strategy for designing a synthetic tumor vaccine: Multi-
component, multivalency and antigen modification. Vaccines (Basel) 2014; 2(3): 549-62.
[http://dx.doi.org/10.3390/vaccines2030549] [PMID: 26344745]

[61]   Milani A, Sangiolo D, Aglietta M, Valabrega G. Recent advances in the development of breast cancer
vaccines. Breast Cancer (Dove Med Press) 2014; 6: 159-68.
[PMID: 25339848]

[62]   Nativi C, Renaudet O. Recent progress in antitumoral synthetic vaccines. ACS Med Chem Lett 2014;
5(11): 1176-8.
[http://dx.doi.org/10.1021/ml5003794] [PMID: 25408824]

[63]   Julien S, Picco G, Sewell R, *et al.* Sialyl-Tn vaccine induces antibody-mediated tumour protection in a
relevant murine model. Br J Cancer 2009; 100(11): 1746-54.
[http://dx.doi.org/10.1038/sj.bjc.6605083] [PMID: 19436292]

[64]   Holmberg LA, Sandmaier BM. Theratope vaccine (STn-KLH). Expert Opin Biol Ther 2001; 1(5):
881-91.
[http://dx.doi.org/10.1517/14712598.1.5.881] [PMID: 11728222]

[65]   Holmberg LA, Sandmaier BM. Vaccination with Theratope (STn-KLH) as treatment for breast cancer.
Expert Rev Vaccines 2004; 3(6): 655-63.
[http://dx.doi.org/10.1586/14760584.3.6.655] [PMID: 15606349]

[66]   Miles D, Roché H, Martin M, Perren TJ, Cameron DA, Glaspy J. Phase III multicenter clinical trial of
the sialyl-TN (STn)-keyhole limpet hemocyanin (KLH) vaccine for metastatic breast cancer.
Oncologist 2011; 16(8): 1092-100.

[67]   Dube DH, Bertozzi CR. Glycans in cancer and inflammation--potential for therapeutics and diagnostics. Nat Rev Drug Discov 2005; 4(6): 477-88.
[http://dx.doi.org/10.1038/nrd1751] [PMID: 15931257]

[68]   PR Newswire. Biomira and Merck KGaA Announce Phase III Theratope(R) Vaccine Trial Does Not Meet Primary Endpoints. 2003. Available from: http://www.prnewswire.com/news-releases/biomira-and-merck-kgaa-announce-phase-iii-theratoper-vaccine-trial-does-not-meet-primary-endpoints-71326627.html

[69]   Ibrahim NK, Murray JL, Zhou D, *et al.* Survival advantage in patients with metastatic breast cancer receiving endocrine therapy plus sialyl Tn-KLH vaccine: post hoc analysis of a large randomized trial. J Cancer 2013; 4(7): 577-84.
[http://dx.doi.org/10.7150/jca.7028] [PMID: 23983823]

[70]   Yin Z, Chowdhury S, McKay C, *et al.* Significant impact of immunogen design on the diversity of antibodies generated by carbohydrate-based anticancer vaccine. ACS Chem Biol 2015; 10(10): 2364-72.
[http://dx.doi.org/10.1021/acschembio.5b00406] [PMID: 26262839]

[71]   Kaiser A, Gaidzik N, Westerlind U, *et al.* A synthetic vaccine consisting of a tumor-associated sialyl-T(N)-MUC1 tandem-repeat glycopeptide and tetanus toxoid: induction of a strong and highly selective immune response. Angew Chem Int Ed Engl 2009; 48(41): 7551-5.
[http://dx.doi.org/10.1002/anie.200902564] [PMID: 19685547]

[72]   Cai H, Palitzsch B, Hartmann S, *et al.* Antibody induction directed against the tumor-associated MUC4 glycoprotein. ChemBioChem 2015; 16(6): 959-67.
[http://dx.doi.org/10.1002/cbic.201402689] [PMID: 25755023]

[73]   Ragupathi G, Gardner JR, Livingston PO, Gin DY. Natural and synthetic saponin adjuvant QS-21 for vaccines against cancer. Expert Rev Vaccines 2011; 10(4): 463-70.
[http://dx.doi.org/10.1586/erv.11.18] [PMID: 21506644]

[74]   Wilson RM, Danishefsky SJ. A vision for vaccines built from fully synthetic tumor-associated antigens: from the laboratory to the clinic. J Am Chem Soc 2013; 135(39): 14462-72.
[http://dx.doi.org/10.1021/ja405932r] [PMID: 23944352]

[75]   Galan MC, Dumy P, Renaudet O. Multivalent glyco(cyclo)peptides. Chem Soc Rev 2013; 42(11): 4599-612.
[http://dx.doi.org/10.1039/C2CS35413F] [PMID: 23263159]

[76]   Johannes M, Reindl M, Gerlitzki B, Schmitt E, Hoffmann-Röder A. Synthesis and biological evaluation of a novel MUC1 glycopeptide conjugate vaccine candidate comprising a 4'-deoxy-4'-fluoro-Thomsen-Friedenreich epitope. Beilstein J Org Chem 2015; 11(1): 155-61.
[http://dx.doi.org/10.3762/bjoc.11.15] [PMID: 25670999]

[77]   Richichi B, Thomas B, Fiore M, *et al.* A cancer therapeutic vaccine based on clustered Tn-antigen mimetics induces strong antibody-mediated protective immunity. Angew Chem Int Ed Engl 2014; 53(44): 11917-20.
[http://dx.doi.org/10.1002/anie.201406897] [PMID: 25168881]

[78]   Hoffmann-Röder A, Kaiser A, Wagner S, *et al.* Synthetic antitumor vaccines from tetanus toxoid

conjugates of MUC1 glycopeptides with the Thomsen-Friedenreich antigen and a fluorine-substituted analogue. Angew Chem Int Ed Engl 2010; 49(45): 8498-503.
[http://dx.doi.org/10.1002/anie.201003810] [PMID: 20878823]

[79]   Yan J, Chen X, Wang F, Cao H. Chemoenzymatic synthesis of mono- and di-fluorinated Thomsen-Friedenreich (T) antigens and their sialylated derivatives. Org Biomol Chem 2013; 11(5): 842-8.
[http://dx.doi.org/10.1039/C2OB26989A] [PMID: 23241945]

[80]   Renaudet O, BenMohamed L, Dasgupta G, Bettahi I, Dumy P. Towards a self-adjuvanting multivalent B and T cell epitope containing synthetic glycolipopeptide cancer vaccine. ChemMedChem 2008; 3(5): 737-41.
[http://dx.doi.org/10.1002/cmdc.200700315] [PMID: 18205167]

[81]   Thompson P, Lakshminarayanan V, Supekar NT, *et al.* Linear synthesis and immunological properties of a fully synthetic vaccine candidate containing a sialylated MUC1 glycopeptide. Chem Commun (Camb) 2015; 51(50): 10214-7.
[http://dx.doi.org/10.1039/C5CC02199E] [PMID: 26022217]

[82]   Scott AM, Allison JP, Wolchok JD. Monoclonal antibodies in cancer therapy. Cancer Immun 2012; 12: 14.
[PMID: 22896759]

[83]   Pucca MB, Bertolini TB, Barbosa JE, Galina SV, Porto GS. Therapeutic monoclonal antibodies: scFv patents as a marker of a new class of potential biopharmaceuticals. Braz J Pharm Sci 2011; 47(1): 31-8.

[84]   Ali M, Hitomi K, Nakano H. Generation of monoclonal antibodies using simplified single-cell reverse transcription-polymerase chain reaction and cell-free protein synthesis. J Biosci Bioeng 2006; 101(3): 284-6.
[http://dx.doi.org/10.1263/jbb.101.284] [PMID: 16716934]

[85]   Babcook JS, Leslie KB, Olsen OA, Salmon RA, Schrader JW. A novel strategy for generating monoclonal antibodies from single, isolated lymphocytes producing antibodies of defined specificities. Proc Natl Acad Sci USA 1996; 93(15): 7843-8.
[http://dx.doi.org/10.1073/pnas.93.15.7843] [PMID: 8755564]

[86]   Smith K, Garman L, Wrammert J, *et al.* Rapid generation of fully human monoclonal antibodies specific to a vaccinating antigen. Nat Protoc 2009; 4(3): 372-84.
[http://dx.doi.org/10.1038/nprot.2009.3] [PMID: 19247287]

[87]   Ecker DM, Jones SD, Levine HL. The therapeutic monoclonal antibody market. MAbs 2015; 7(1): 9-14.
[http://dx.doi.org/10.4161/19420862.2015.989042] [PMID: 25529996]

[88]   Liu JK. The history of monoclonal antibody development - Progress, remaining challenges and future innovations. Ann Med Surg (Lond) 2014; 3(4): 113-6.
[http://dx.doi.org/10.1016/j.amsu.2014.09.001] [PMID: 25568796]

[89]   Chames P, Van Regenmortel M, Weiss E, Baty D. Therapeutic antibodies: successes, limitations and hopes for the future. Br J Pharmacol 2009; 157(2): 220-33.
[http://dx.doi.org/10.1111/j.1476-5381.2009.00190.x] [PMID: 19459844]

[90]   Kubota T, Niwa R, Satoh M, Akinaga S, Shitara K, Hanai N. Engineered therapeutic antibodies with improved effector functions. Cancer Sci 2009; 100(9): 1566-72.
[http://dx.doi.org/10.1111/j.1349-7006.2009.01222.x] [PMID: 19538497]

[91]   Carter P. Improving the efficacy of antibody-based cancer therapies. Nat Rev Cancer 2001; 1(2): 118-29.
[http://dx.doi.org/10.1038/35101072] [PMID: 11905803]

[92]   Brekke OH, Sandlie I. Therapeutic antibodies for human diseases at the dawn of the twenty-first century. Nat Rev Drug Discov 2003; 2(1): 52-62.
[http://dx.doi.org/10.1038/nrd984] [PMID: 12509759]

[93]   Ludwig DL, Pereira DS, Zhu Z, Hicklin DJ, Bohlen P. Monoclonal antibody therapeutics and apoptosis. Oncogene 2003; 22(56): 9097-106.
[http://dx.doi.org/10.1038/sj.onc.1207104] [PMID: 14663488]

[94]   Scott AM, Wolchok JD, Old LJ. Antibody therapy of cancer. Nat Rev Cancer 2012; 12(4): 278-87.
[http://dx.doi.org/10.1038/nrc3236] [PMID: 22437872]

[95]   Kim EG, Kim KM. Strategies and advancement in antibody-drug conjugate optimization for targeted cancer therapeutics. Biomol Ther (Seoul) 2015; 23(6): 493-509.
[http://dx.doi.org/10.4062/biomolther.2015.116] [PMID: 26535074]

[96]   Gerber H-P, Sapra P, Loganzo F, May C. Combining antibody-drug conjugates and immune-mediated cancer therapy: What to expect? Biochem Pharmacol 2016; 102: 1-6.
[http://dx.doi.org/10.1016/j.bcp.2015.12.008] [PMID: 26686577]

[97]   Zhu Y, Choi SH, Shah K. Multifunctional receptor-targeting antibodies for cancer therapy. Lancet Oncol 2015; 16(15): e543-54.
[http://dx.doi.org/10.1016/S1470-2045(15)00039-X] [PMID: 26545843]

[98]   Thakur A, Lum LG. "NextGen" biologics: Bispecific antibodies and emerging clinical results. Expert Opin Biol Ther 2016; 16(5): 675-88.
[http://dx.doi.org/10.1517/14712598.2016.1150996] [PMID: 26848610]

[99]   Cunto-Amesty G, Luo P, Monzavi-Karbassi B, Lees A, Kieber-Emmons T. Exploiting molecular mimicry to broaden the immune response to carbohydrate antigens for vaccine development. Vaccine 2001; 19(17-19): 2361-8.
[http://dx.doi.org/10.1016/S0264-410X(00)00527-2] [PMID: 11257361]

[100]   Rittenhouse-Diakun K, Xia Z, Pickhardt D, Morey S, Baek M-G, Roy R. Development and characterization of monoclonal antibody to T-antigen: (gal beta1-3GalNAc-alpha-O). Hybridoma 1998; 17(2): 165-73.
[http://dx.doi.org/10.1089/hyb.1998.17.165] [PMID: 9627057]

[101]   Ferguson K, Yadav A, Morey S, *et al.* Preclinical studies with JAA-F11 anti-Thomsen-Friedenreich monoclonal antibody for human breast cancer. Future Oncol 2014; 10(3): 385-99.
[http://dx.doi.org/10.2217/fon.13.209] [PMID: 24559446]

[102]   Karsten U, Butschak G, Cao Y, Goletz S, Hanisch F-G. A new monoclonal antibody (A78-G/A7) to the Thomsen-Friedenreich pan-tumor antigen. Hybridoma 1995; 14(1): 37-44.
[http://dx.doi.org/10.1089/hyb.1995.14.37] [PMID: 7539400]

[103] Kjeldsen T, Clausen H, Hirohashi S, Ogawa T, Iijima H, Hakomori S. Preparation and characterization of monoclonal antibodies directed to the tumor-associated O-linked sialosyl-2----6 alpha-N-acetylgala-ctosaminyl (sialosyl-Tn) epitope. Cancer Res 1988; 48(8): 2214-20.
[PMID: 2450649]

[104] Matsumoto-Takasaki A, Horie J, Sakai K, *et al.* Isolation and characterization of anti-T-antigen single chain antibodies from a phage library. Biosci Trends 2009; 3(3): 87-95.
[PMID: 20103829]

[105] Dahlenborg K, Hultman L, Carlsson R, Jansson B. Human monoclonal antibodies specific for the tumour associated Thomsen-Friedenreich antigen. Int J Cancer 1997; 70(1): 63-71.
[http://dx.doi.org/10.1002/(SICI)1097-0215(19970106)70:1<63::AID-IJC10>3.0.CO;2-E] [PMID: 8985092]

[106] Tarp MA, Sørensen AL, Mandel U, *et al.* Identification of a novel cancer-specific immunodominant glycopeptide epitope in the MUC1 tandem repeat. Glycobiology 2007; 17(2): 197-209.
[http://dx.doi.org/10.1093/glycob/cwl061] [PMID: 17050588]

[107] Sørensen AL, Reis CA, Tarp MA, *et al.* Chemoenzymatically synthesized multimeric Tn/STn MUC1 glycopeptides elicit cancer-specific anti-MUC1 antibody responses and override tolerance. Glycobiology 2006; 16(2): 96-107.
[http://dx.doi.org/10.1093/glycob/cwj044] [PMID: 16207894]

[108] Thurnher M, Clausen H, Sharon N, Berger EG. Use of O-glycosylation-defective human lymphoid cell lines and flow cytometry to delineate the specificity of Moluccella laevis lectin and monoclonal antibody 5F4 for the Tn antigen (GalNAc α 1-O-Ser/Thr). Immunol Lett 1993; 36(3): 239-43.
[http://dx.doi.org/10.1016/0165-2478(93)90095-J] [PMID: 8370596]

[109] Pancino GF, Osinaga E, Vorauher W, *et al.* Production of a monoclonal antibody as immunohistochemical marker on paraffin embedded tissues using a new immunization method. Hybridoma 1990; 9(4): 389-95.
[http://dx.doi.org/10.1089/hyb.1990.9.389] [PMID: 2210779]

[110] Terasawa K, Furumoto H, Kamada M, Aono T. Expression of Tn and sialyl-Tn antigens in the neoplastic transformation of uterine cervical epithelial cells. Cancer Res 1996; 56(9): 2229-32.
[PMID: 8616877]

[111] Numata Y, Nakada H, Fukui S, *et al.* A monoclonal antibody directed to Tn antigen. Biochem Biophys Res Commun 1990; 170(3): 981-5.
[http://dx.doi.org/10.1016/0006-291X(90)90488-9] [PMID: 2390097]

[112] Ando H, Matsushita T, Wakitani M, *et al.* Mouse-human chimeric anti-Tn IgG1 induced anti-tumor activity against Jurkat cells *in vitro* and *in vivo*. Biol Pharm Bull 2008; 31(9): 1739-44.
[http://dx.doi.org/10.1248/bpb.31.1739] [PMID: 18758069]

[113] King MJ, Parsons SF, Wu AM, Jones N. Immunochemical studies on the differential binding properties of two monoclonal antibodies reacting with Tn red cells. Transfusion 1991; 31(2): 142-9.
[http://dx.doi.org/10.1046/j.1537-2995.1991.31291142945.x] [PMID: 1847560]

[114] Danussi C, Coslovi A, Campa C, *et al.* A newly generated functional antibody identifies Tn antigen as a novel determinant in the cancer cell-lymphatic endothelium interaction. Glycobiology 2009; 19(10):

1056-67.
[http://dx.doi.org/10.1093/glycob/cwp085] [PMID: 19528665]

[115] Welinder C, Baldetorp B, Borrebaeck C, Fredlund BM, Jansson B. A new murine IgG1 anti-Tn monoclonal antibody with *in vivo* anti-tumor activity. Glycobiology 2011; 21(8): 1097-107.
[http://dx.doi.org/10.1093/glycob/cwr048] [PMID: 21470982]

[116] Colcher D, Hand PH, Nuti M, Schlom J. A spectrum of monoclonal antibodies reactive with human mammary tumor cells. Proc Natl Acad Sci USA 1981; 78(5): 3199-203.
[http://dx.doi.org/10.1073/pnas.78.5.3199] [PMID: 6789331]

[117] Pant KD, Jain A, McCracken JD, Thompson K. Immunohistochemical examination of anti-STn monoclonal antibodies LLU9B4, B72.3, and B35.2 for their potential use as tumor markers. Dig Dis Sci 2008; 53(8): 2189-94.
[http://dx.doi.org/10.1007/s10620-007-0137-2] [PMID: 18299983]

[118] An Y, Han W, Chen X, *et al.* A novel anti-sTn monoclonal antibody 3P9 Inhibits human xenografted colorectal carcinomas. J Immunother 2013; 36(1): 20-8.
[http://dx.doi.org/10.1097/CJI.0b013e31827810d1] [PMID: 23211624]

[119] Muraro R, Kuroki M, Wunderlich D, *et al.* Generation and characterization of B72.3 second generation monoclonal antibodies reactive with the tumor-associated glycoprotein 72 antigen. Cancer Res 1988; 48(16): 4588-96.
[PMID: 3396010]

[120] Kurosaka A, Fukui S, Kitagawa H, *et al.* Mucin-carbohydrate directed monoclonal antibody. FEBS Lett 1987; 215(1): 137-9.
[http://dx.doi.org/10.1016/0014-5793(87)80128-X] [PMID: 3569536]

[121] Yuasa N, Koyama T, Subedi GP, Yamaguchi Y, Matsushita M, Fujita-Yamaguchi Y. Expression and structural characterization of anti-T-antigen single-chain antibodies (scFvs) and analysis of their binding to T-antigen by surface plasmon resonance and NMR spectroscopy. J Biochem 2013; 154(6): 521-9.
[http://dx.doi.org/10.1093/jb/mvt089] [PMID: 24098012]

[122] Blixt O, Boos I, Mandel U. Glycan microarray analysis of tumor-associated antibodies. In: Kosma P, Müller-Loennies S, Eds. Anticarbohydrate antibodies: From molecular basis to clinical application. Vienna: Springer 2012; pp. 283-306.
[http://dx.doi.org/10.1007/978-3-7091-0870-3_12]

[123] Zamri N, Masuda N, Oura F, Yajima Y, Nakada H, Fujita-Yamaguchi Y. Effects of two monoclonal antibodies, MLS128 against Tn-antigen and 1H7 against insulin-like growth factor-I receptor, on the growth of colon cancer cells. Biosci Trends 2012; 6(6): 303-12.
[PMID: 23337790]

[124] Zhang S, Walberg LA, Ogata S, *et al.* Immune sera and monoclonal antibodies define two configurations for the sialyl Tn tumor antigen. Cancer Res 1995; 55(15): 3364-8.
[PMID: 7614472]

[125] Tarp MA, Sørensen AL, Mandel U, *et al.* Identification of a novel cancer-specific immunodominant glycopeptide epitope in the MUC1 tandem repeat. Glycobiology 2007; 17(2): 197-209.
[http://dx.doi.org/10.1093/glycob/cwl061] [PMID: 17050588]

CHAPTER 2

Cell Surface Glycans as Viral Entry Factors and Targets for Broadly Acting Antivirals

Che C. Colpitts[1,2,*] and Thomas F. Baumert[1,2,3,*]

[1] *Inserm, U1110, Institut de Recherche sur les Maladies Virales et Hépatiques, 67000 Strasbourg, France*

[2] *Université de Strasbourg, 67000 Strasbourg, France*

[3] *Institut Hospitalo-Universitaire, Pôle Hépato-digestif, Hôpitaux Universitaires de Strasbourg, 67000 Strasbourg, France.*

Abstract: Cellular glycans play key roles in the infection process of many human viruses. Viral attachment to heparan sulfate (HS) or sialic acid (SA) moieties in cell surface glycans is a critical and conserved step for the entry of many human viruses, including clinically important human pathogens such as hepatitis B virus, hepatitis C virus, human immunodeficiency virus and influenza virus. As such, glycans are attractive targets for broadly acting antivirals. Molecules that mimic HS or SA interfere with viral attachment by competing for binding of virion glycoproteins to cellular glycans. Modulation of the levels of cellular glycans also affects viral attachment. These approaches show great promise based on their broad-spectrum activities, but the molecules identified so far often possess undesirable pharmacological properties resulting in potential adverse effects. Here, we describe the mechanisms involved in glycan binding, discuss broadly acting glycan-targeted antiviral strategies, and provide perspectives for the rational design of broad-spectrum small molecule entry inhibitors with broad-spectrum activities and appropriate pharmacological properties.

Keywords: Antivirals, Broad-spectrum antiviral therapy, Chemical biology, Glycan mimetics, Glycosaminoglycans, Glycotherapy, Heparan sulfate proteoglycans, Heparin, Hepatitis B virus, Hepatitis C virus, Herpes simplex virus,

---

** **Corresponding authors Thomas F. Baumert:** Université de Strasbourg, 67000 Strasbourg, France; Tel: +33 3 68 85 37 03; Fax: +33 3 68 85 37 24; E-mail: thomas.baumert@unistra.fr and **Che C. Colpitts:** Université de Strasbourg, 67000 Strasbourg, France; Tel: +33 3 68 85 37 38; Fax: +33 3 68 85 37 24; E-mail: colpitts@unistra.fr*

Human immunodeficiency virus, Influenza virus, Polyphenols, Sialic acid, Sialoglycans, Small molecule entry inhibitors, Viral attachment, Viral entry, Virus-host interactions.

INTRODUCTION

Cell surface glycans mediate the initial infection steps of many pathogens, including most human viruses. An overwhelming majority of these viruses initiate infection of their target cells by low-affinity binding to cell surface glycans, such as glycosaminoglycans (GAGs) or sialoglycans (SGs). GAGs and SGs are ubiquitous on cells, and many microbes have evolved to exploit them for initial attachment to their target cells [1]. Among the GAGs, heparin/heparan sulfate (HS) is the most commonly used viral attachment site. The high density of cell-surface HS allows microbes to bind and increase their concentration at the cell surface to facilitate subsequent entry and infection steps. Other viruses recognize and bind to sialic acid (SA), which is abundantly expressed in SGs on the cell surface [2]. These interactions with glycans are critical to capture virions from extracellular spaces and concentrate them in the vicinity of other entry factors and signalling molecules to allow high affinity binding and uptake into the cell. As such, glycans represent attractive targets for broad-spectrum antiviral therapies.

GLYCAN STRUCTURE

GAGs are unbranched polysaccharides comprised of repeating disaccharide units of an amino sugar (*N*-acetylglucosamine or *N*-acetylgalactosamine) and either a uronic sugar (either glucuronic acid or iduronic acid) or galactose [3] and usually linked to proteins *via O*-linked or *N*-linked glycosylation. GAGs are classified into four groups that reflect the composition of the disaccharide unit. These include (HS), chondroitin/dermatan sulfate, keratan sulfate or hyaluronic acid [4]. The repeating disaccharide unit of GAGs is variably sulfated [5], leading to high diversity and complexity and allowing for specific binding with distinct GAG-binding proteins. Indeed, interactions between GAGs and a diverse spectrum of GAG-binding proteins regulate many biological processes, including cell growth and proliferation pathways, cell adhesion or migration, and tissue hydration [4]. GAG-binding proteins interact with negatively charged sulfates and carboxylates in GAGs *via* a binding pocket of positively charged basic amino acids. A consensus sequence of XBBXBX or XBBBXXBX (where B is a basic lysine or arginine residue) is typically found in proteins that bind to HS [6]. Binding affinity depends on the overall shape and conformation of GAGs and is affected by the number and orientation of the sulfate group charges [3].

SGs are a distinct group of cellular glycans, comprised of (SA) attached to the termini of *N*-linked and *O*-linked glycans. The carboxylate group at the 1-carbon position of SA is ionized at physiological pH, allowing for interactions with basic residues of saccharide-binding proteins such as lectins [7]. Binding between SGs and SG-binding proteins is predominantly mediated by extensive hydrogen bonding between the carboxylate, hydroxyls and N-acetyl group of SA and polar amino acid residues.

HEPARAN SULFATE-BINDING VIRUSES

A large and diverse group of human viruses bind to HS proteoglycans, *via* interactions between viral glycoproteins and negatively charged HS moieties in cellular GAGs. These viruses are summarized in Table **1**. Most hepatotropic viruses such as hepatitis C virus (HCV) [8 - 10], hepatitis B virus (HBV) and its satellite, hepatitis D virus [11 - 14], as well as hepatitis E virus [15], require a primary attachment step to HS. Moreover, other viruses that infect a wide range of cell types similarly require HS for binding. These include viruses with RNA genomes, such as human immunodeficiency virus (HIV) [16], dengue virus [17], filoviruses [18], Sindbis virus (SINV) [19], respiratory syncytial virus (RSV) [20 - 22] and other paramyxoviruses [23, 24], Rift Valley fever virus [25], severe acute respiratory syndrome-associated coronavirus [26], as well as DNA viruses, including herpes simplex virus 1 and 2 (HSV-1/-2) [27 - 30], human cytomegalovirus (CMV) [31], varicella zoster virus [32], human herpesvirus 8 [33, 34], vaccinia virus (VACV) [35], adenovirus (AdV) types 2 and 5 [36, 37], some strains of norovirus [38] and human papillomavirus (HPV) [39]. Merkel cell virus initially interacts with HS-containing GAGs, although SA is also required [40]. Other viruses, including the rhabdovirus vesicular stomatitis virus (VSV), are thought to also require HS for binding [41], although the specific details of the interactions remain unclear.

Table 1. Viral attachment to cellular glycans. The majority of human viruses bind to heparan sulfate (HS) or sialic acid (SA).

Virus family	Virus	Glycan	Reference
Adenoviridae	Adenovirus types 2 and 5	HS	[36, 37]
	Adenovirus type 37	SA	[66, 67]
Bunyaviridae	Rift Valley fever virus	HS	[25]
Caliciviridae	Norovirus genogroup (G) II	HS	[38]
	Norovirus GII.3 and GII.4	SA	[58]
Coronaviridae	SARS-coronavirus	HS	[26]
	Coronavirus OC43	SA	[56]
Flaviviridae	Hepatitis C virus	HS	[8-10]
	Dengue virus	HS	[17]
Filoviridae	Ebola virus	HS	[18]
Hepadnaviridae	Hepatitis B virus (and delta virus)	HS	[11-14]
Hepaviridae	Hepatitis E virus	HS	[15]
Herpesviridae	Herpes simplex virus type 1	HS	[27-30]
	Herpes simplex virus type 2	HS	[27-30]
	Human cytomegalovirus	HS	[31]
	Varicella zoster virus	HS	[32]
	Human herpesvirus 8	HS	[33, 34]
Orthomyxoviridae	Influenza A virus	SA	[53]
	Influenza B virus	SA	[54]
	Influenza C virus	SA	[55]
Papillomaviridae	Human papillomavirus types 16 and 33	HS	[39]
Paramyxoviridae	Respiratory syncytial virus	HS	[20-22]
	Hendra virus	HS	[23]
	Nipah virus	HS	[23]
	Parainfluenza	HS	[24]
	Sendai virus	SA	[57]
Picornaviridae	Coxsackievirus A24	SA	[65]
	Enterovirus 70	SA	[60]
Polyomaviridae	JC polyomavirus	SA	[61]
	BK virus	SA	[62]
	Merkel cell virus	SA/HS	[40]
Poxviridae	Vaccinia virus	HS	[35]
Reoviridae	Reovirus type 3	SA	[59]
Retroviridae	Human immunodeficiency virus	HS	[16]
Togaviridae	Sindbis virus	HS	[19]

Viral attachment to HS moieties in cellular glycans is highly conserved, requiring interactions between binding pockets of basic amino acids in the virion glycoproteins and negatively charged HS residues. HSV-1 primary attachment is perhaps the best characterized among the HS-binding viruses. HSV-1 glycoprotein B (gB) and glycoprotein C (gC) are responsible for primary attachment to cellular HS [28, 42, 43] *via* electrostatic interactions between basic residues in gB/gC and the negatively charged sulfate esters and carboxylate groups of HS. A basic lysine-rich region of gB from residue 68 to 76 (KPKKNKKPK) is required for binding to heparin and HS [44]. Glycoprotein gC also has basic residues involved in HS binding, including Arg-143, Arg-145 and Arg-147 [45] as well as Arg-129, Arg-130, Arg-151, Arg-155 and Arg-160 [46]. Non-ionic hydrophobic residues such as Thr-150 and Ile-142 are apparently also involved [45, 46] and may contribute to the binding energy or correct positioning of the basic residues.

Similar heparin-binding domains are found in other virion glycoproteins. Analysis of over 1500 non-redundant sequences identified conserved basic residues at specific positions within the N-terminal 27-amino-acid hypervariable region (HVR1) of the HCV E2 protein [47]. These residues were proposed to interact with negatively charged cell surface GAGs. Indeed, E2 with a deletion of the basic HVR-1 region showed a decreased ability to bind to heparin [8], demonstrating the importance of these residues. Likewise, HIV glycoprotein gp120 contains four heparin-binding domains of basic amino acids: [166]RGKVQK[171], [304]RRKIR[308], [500]KAKRR[504] [48], and a binding pocket of Lys-121, Arg-419, Lys-421 and Lys-432 [49]. VACV A27 protein binds to HS through a lysine-rich binding domain ([21]STKAAKKPEAKR[32]) [50]. The RSV G protein has a lysine-rich heparin-binding domain ([184]AICKRIPNKKPGKKT[198] or [183]KSICKTIPSNKPKKK[197], depending on the subgroup) [22]. Nonenveloped viruses also bind to HS through similar interactions. A conserved region of basic amino acids in the HPV L1 protein the consensus sequence from nine HPV types is XBBBBXB, where B is Lys, Arg or His is involved in binding to heparin and HS [51]. The [91]KKTK[94] domain of the AdV fiber shaft protein has also been implicated in heparin binding [36]. Therefore, binding between most virion glycoproteins and cellular GAGs is mediated by very similar ionic interactions between the same basic amino acids and negatively charged HS moieties.

SIALIC ACID-BINDING VIRUSES

SA was the first virus receptor identified, in the context of influenza infection [52]. Since that discovery, SA was shown to be a receptor for a diverse group of viruses and other pathogens [2]. Viruses in this group include influenza A virus (IAV) [53], influenza B virus [54], influenza C virus [55], some human coronavirus strains [56], some paramyxoviruses such as Sendai virus [57], some strains of norovirus [58], reovirus [59], enterovirus 70 [60], JC polyomavirus [61], BK virus [62] and others [63], as shown in Table 1. Additionally, certain strains and isolates of human rhinovirus, coxsackievirus and AdV bind to sialic acids [64 - 67]. Merkel cell virus is thought to bind to SA as a secondary binding step following initial interaction with HS [40]. Similar to HS-binding viruses, attachment of SA-binding viruses requires low-affinity interactions between binding pockets in virion glycoproteins and SA moieties in cellular SGs.

Proteins that bind to SA, such as IAV hemagglutinin (HA), do so by hydrogen bonding between polar amino acids and the functional groups of SA (a carboxylate at C1, a hydroxyl at C2, an N-acetyl group at C5 and a glycerol group at C6). The conserved residues in the SA-binding site of IAV HA are Tyr-98, Trp-153, Glu-190, Leu-194 and His-183 [68]. The pyranose core of SA rests on top of the aromatic residues, Tyr-98 and Trp-153. The carboxylate group of SA forms hydrogen bonds with Ser-136, and to the amide of the Asn-137 peptide bond.

The hydroxyl groups in the glycerol chain of SA hydrogen bond to His-183, Glu-190, and Tyr-98 [68]. The N-acetyl group of SA also forms hydrogen bonds (to the carbonyl of the Gly-135 peptide bond). Moreover, van der Waals interactions (such as between the methyl of the N-acetyl group and Trp-153) are also involved in the binding.

Similar hydrogen bonding interactions are described for other viral proteins that bind SA. For example, the hemagglutinin of the paramyxovirus Newcastle disease virus binds to SA by hydrogen bonding mediated by amino acids Glu-401, Arg-416 and Tyr-526 [69]. The binding pocket also contains Arg-498, Ser-418, Tyr-

317, Glu-258 and Ser-237 [69], all of which can form hydrogen bonds. Reovirus sigma 1 has a binding pocket consisting of Asn-198, Arg-202, Leu-203, Pro-204 and Gly-205, which interact with sialic acid through a similar network of hydrogen bonds and van der Waals interactions [70].

GLYCANS AS ANTIVIRAL TARGETS

The highly conserved binding mechanisms point to viral attachment as a promising broad-spectrum antiviral target. Indeed, heparin/HS and appropriately shaped charged molecules that mimic HS disrupt the binding of a broad group of viruses. Similarly, molecules that mimic SA compete for receptor binding to inhibit attachment of SA-dependent viruses. One major advantage of glycans as an antiviral target is the broad spectrum of viruses that could potentially be targeted, which could be useful for co-infected patients or in the context of emerging viral infections. Nonetheless, major limitations still hamper glycan-targeted antiviral therapies, including poor pharmacological properties of the molecules and off-target effects. Furthermore, the individual interactions between a single binding pocket and HS or SA are weak. Receptor binding is cooperative and likely depends on multivalent interactions between virion glycoproteins and multiple glycan moieties. Consequently, monovalent receptor mimetics are unable to effectively compete with the multivalent interactions of the virions with the cells [71], whereas glycopolymers and other polyvalent glycan mimetics are likely to be more effective competitors.

Inhibitors of HS-Binding Viruses

It has long been known that soluble heparin, HS mimetics, and other polyanions inhibit the infectivity of HS-dependent viruses. Indeed, sulfated polysaccharide derivatives and sulfated polymers mimic HS and compete for the binding of virion glycoproteins. Dextran sulphate, pentosan polysulfate and other sulfated polysaccharides inhibit the infectivity of HSV-1/-2, CMV, VSV, SINV, HIV and some flaviviruses [72, 73], all of which bind to HS. Similarly, many polysulfonates and sulfonic acid polymers inhibit the infectivity of the HS-binding HIV [6, 74, 75] and the polysulfonate suramin inhibits adsorption of the HS-binding HSV-1 [76]. Sulfated homologues of heparin also inhibit HCV entry [77].

Collectively, these compounds block the binding of positively charged amino acids residues in the viral gp120 to the cell surface HS [78]. Polycarboxylates such as aurintricarboxylic acid [79], as well as polyhydroxycarboxylates derived from phenolic compounds [80, 81], also have inhibitory activity against HS-binding viruses. It is likely that the polycarboxylates also disrupt the interaction between basic regions of virion glycoproteins and cellular HS [82]. Interestingly, synthetic anti-lipopolysaccharide (LPS) peptides (SALPs) were shown to bind to HS moieties on the cell surface and block entry of a variety of HS-binding viruses, including HIV-1, HBV, HCV and HSV-1/-2 [83].

Despite more than 30 years of research, there has so far been limited success to identify a small-molecule viral attachment inhibitor with pharmacological properties suitable for clinical use. HS mimetics suffer from drawbacks that limit their use *in vivo*. Polysulfates, for example, are poorly absorbed after oral administration [84], and cause thrombocytopenia when administered intravenously [85]. Most sulfated polymers have strong anticoagulant activity [86], limiting their utility for antiviral therapy. It was possible, however, to dissociate the antiviral effects of sulfated polysaccharides from their antithrombin activity [87, 88]. The number, distribution and spatial configuration of the sulphate groups were proposed to differentially influence the antiviral and antithrombin activities. Therefore, appropriate chemical modifications or novel compound scaffolds could help to overcome the limitations of HS mimetic antivirals.

Many natural products of different scaffolds are biologically active, and may not suffer from similar limitations. Indeed, some natural products inhibit viral attachment to cellular glycans. Sulfated polysaccharides from various species of algae [89, 90], and natural products fucoidan and carrageenan [72] inhibit the infectivity of various HS-binding enveloped viruses. Recently, two tannins (chebulagic acid and punicalagin) isolated from the *Terminalia chebula* tree were shown to block attachment of a diverse group of viruses that utilize GAGs for binding [91]. In the case of HSV-1, chebulagic acid and punicalagin prevented HSV-1 glycoproteins from interacting with cell surface GAGs [92].

Inhibitors of SA-Binding Viruses

SA derivatives and sialyl mimetics are active against SA-binding viruses, such as IAV, but do not inhibit binding of HS-dependent viruses. Polyvalent inhibitors are far more potent than monovalent ones, given that binding of virion glycoproteins such as IAV HA to soluble sialyl mimetics requires multiple HA-sialic acid contacts [71, 93] and depends on the spatial orientation, size and flexibility of the inhibitor. For example, equine α-macroglobulin with multiple SA residues in appropriate orientations has a million-fold higher potency against IAV than free N-linked oligosaccharides [94]. Optimized synthetic sialylglycopolymers on a poly[N-acryoyloxysuccinimide], polyacrylic acid or polyacrylamide backbone, with variable arrangement and number of sialyl residues, were even more potent against IAV binding, with EC_{50} in the nanomolar range [95, 96]. Similarly, sialic acid-conjugated dendritic polymers and glycopolymers are more effective than monomeric SA at inhibiting IAV binding [97, 98]. Glycopeptides that interfere with IAV binding have also been described. A trivalent glycopeptide SA mimetic, comprised of three sialyl moieties linked by peptide regions, was designed to bind the receptor sites in each monomer of the HA homotrimer [99]. The glycopeptide had high binding affinity for HA, although its anti-IAV activities remain to be determined. Sialyl mimetics should selectively interact with HA and not with the viral neuraminidase (NA) [100, 101], a glycoside hydrolase which cleaves the glycosidic linkages of cellular SA residues. Viral NA may reduce the effectiveness of SA mimetic entry inhibitors as a result of cleavage. However, it may be possible to inhibit both HA and NA with a single molecule [102].

An alternative but complementary method to block primary attachment is to modulate the levels of glycan receptors on the cell surface. On one hand, the removal of glycans from the cell surface can block viral attachment. DAS181, a sialidase from *Actinomyces viscosus* is in phase II clinical development [103]. The sialidase is fused to an epidermal growth factor-like domain to target the sialidase to respiratory epithelial cells [104]. By removing cellular SA, DAS181 prevents attachment of respiratory viruses that utilize SA as a receptor [105, 106]. On the other hand, IAV infection depends on NA activity to release newly produced virions from the surface of infected cells. Oseltamivir (TamiFlu) and zanamivir, clinically approved drugs for IAV treatment, competitively inhibit SA cleavage by

NA, thus inhibiting the spread of IAV particles.

Inhibitors of Both HS- and SA-Binding Viruses

For the most part, molecules targeting HS-binding viruses are not active against SA-binding viruses, and *vice versa*. However, the green tea polyphenol epigallocatechin gallate (EGCG) has a number of interesting biological properties, including antiviral effects against many unrelated enveloped and nonenveloped viruses [107]. Indeed, EGCG has been shown to inhibit a wide spectrum of both HS- and SA-binding viruses, including HIV, IAV, AdV, HBV, HCV, and HSV-1/-2, among others [107 - 117]. Strikingly, EGCG interferes with the primary attachment of unrelated HS-binding and SA-binding viruses [118]. In the case of IAV, modeling studies support the binding of EGCG (and structural analogs) to the SA-binding domain of HA, through hydrogen bonding and σ-π stacking interactions [119], and similar mechanisms are likely involved for HS-binding viruses. Thus, EGCG represents a truly broad-spectrum inhibitor of viral attachment. The structural moieties of EGCG may be uniquely positioned to interact with basic or polar residues in the binding pockets of both HS- and SA-binding virion glycoproteins [118]. A structurally similar natural product, pentagalloyl glucose, inhibited infection by IAV (an SA-binding virus) and HBV (an HS-binding virus) [120, 121], likely by preventing virion binding (at least in the case of IAV) [120]. However, these molecules are unstable in aqueous solutions, poorly absorbed, and undergo metabolic alterations such as oxidation [122], which likely limits their antiviral activity *in vivo*. Thus, these unfavourable pharmacological properties of these molecules preclude their therapeutic use, although it may be possible to rationally design broadly acting attachment inhibitors with improved pharmacology.

PERSPECTIVES

Viral attachment to cellular glycans is a conserved and required process during the entry of many unrelated viruses. As such, it represents an attractive target for broad-spectrum antivirals. Unfortunately, the clinical use of many molecules that target glycan attachment is limited by their pharmacological properties associated with adverse effects. However, identification and characterization of the mechanisms involved in glycan binding opens perspectives for the rational design of small molecule entry inhibitors with broad-spectrum activities, appropriate pharmacological properties and absent adverse effects.

CONFLICT OF INTEREST

The authors confirm that they have no conflict of interest to declare for this publication.

ACKNOWLEDGEMENTS

T.F.B. is supported by funding from the European Union (ERC-2008-A-G-HEPCENT, ERC-2014-AdG-HEPCIR, FP7 HepaMAb, H2020 HEPCAR and Interreg IV FEDER-Hepato-Regio-Net 2012), the Agence Nationale de Recherches sur le SIDA (ANRS), the Direction Générale de l'Offre de Soins (A12027MS), Inserm and the University of Strasbourg Foundation. This work has been published under the framework of the LABEX ANR-10-LABX-0028_HEPSYS and benefits from funding from the state managed by the French National Research Agency as part of the Investments for the future program. C.C.C. acknowledges fellowships from the Canadian Institutes of Health Research (201411MFE-338606-245517) and the Canadian Network on Hepatitis C.

REFERENCES

[1]   Chen Y, Götte M, Liu J, Park PW. Microbial subversion of heparan sulfate proteoglycans. Mol Cells 2008; 26(5): 415-26.
[PMID: 18799929]

[2]   Matrosovich M, Herrler G, Klenk HD. Sialic acid receptors of viruses. Top Curr Chem 2015; 367: 1-28.
[http://dx.doi.org/10.1007/128_2013_466] [PMID: 23873408]

[3]   Raman R, Sasisekharan V, Sasisekharan R. Structural insights into biological roles of protein-glycosaminoglycan interactions. Chem Biol 2005; 12(3): 267-77.
[http://dx.doi.org/10.1016/j.chembiol.2004.11.020] [PMID: 15797210]

[4]   Esko JD, Kimata K, Lindahl U. Proteoglycans and sulfated glycosaminoglycans. In: Varki A, Cummings RD, Esko JD, Freeze HH, Stanley P, Bertozzi CR, Eds. Essentials of Glycobiology. 2nd ed., Cold Spring Harbor, NY 2009.

[5]   David G. Integral membrane heparan sulfate proteoglycans. FASEB J 1993; 7(11): 1023-30.
[PMID: 8370471]

[6]   Cardin AD, Weintraub HJ. Molecular modeling of protein-glycosaminoglycan interactions.

Arteriosclerosis 1989; 9(1): 21-32.
[http://dx.doi.org/10.1161/01.ATV.9.1.21] [PMID: 2463827]

[7]   Varki A. Selectin ligands. Proc Natl Acad Sci USA 1994; 91(16): 7390-7.
        [http://dx.doi.org/10.1073/pnas.91.16.7390] [PMID: 7519775]

[8]   Barth H, Schafer C, Adah MI, *et al.* Cellular binding of hepatitis C virus envelope glycoprotein E2
        requires cell surface heparan sulfate. J Biol Chem 2003; 278(42): 41003-12.
        [http://dx.doi.org/10.1074/jbc.M302267200] [PMID: 12867431]

[9]   Barth H, Schnober EK, Zhang F, *et al.* Viral and cellular determinants of the hepatitis C virus
        envelope-heparan sulfate interaction. J Virol 2006; 80(21): 10579-90.
        [http://dx.doi.org/10.1128/JVI.00941-06] [PMID: 16928753]

[10]   Morikawa K, Zhao Z, Date T, *et al.* The roles of CD81 and glycosaminoglycans in the adsorption and
        uptake of infectious HCV particles. J Med Virol 2007; 79(6): 714-23.
        [http://dx.doi.org/10.1002/jmv.20842] [PMID: 17457918]

[11]   Schulze A, Gripon P, Urban S. Hepatitis B virus infection initiates with a large surface protein-
        dependent binding to heparan sulfate proteoglycans. Hepatology 2007; 46(6): 1759-68.
        [http://dx.doi.org/10.1002/hep.21896] [PMID: 18046710]

[12]   Leistner CM, Gruen-Bernhard S, Glebe D. Role of glycosaminoglycans for binding and infection of
        hepatitis B virus. Cell Microbiol 2008; 10(1): 122-33.
        [PMID: 18086046]

[13]   Lamas Longarela O, Schmidt TT, Schöneweis K, *et al.* Proteoglycans act as cellular hepatitis delta
        virus attachment receptors. PLoS One 2013; 8(3): e58340.
        [http://dx.doi.org/10.1371/journal.pone.0058340] [PMID: 23505490]

[14]   Verrier ER, Colpitts CC, Bach C, *et al.* A targeted functional RNA interference screen uncovers
        glypican 5 as an entry factor for hepatitis B and D viruses. Hepatology 2016; 63(1): 35-48.
        [http://dx.doi.org/10.1002/hep.28013] [PMID: 26224662]

[15]   Kalia M, Chandra V, Rahman SA, Sehgal D, Jameel S. Heparan sulfate proteoglycans are required for
        cellular binding of the hepatitis E virus ORF2 capsid protein and for viral infection. J Virol 2009;
        83(24): 12714-24.
        [http://dx.doi.org/10.1128/JVI.00717-09] [PMID: 19812150]

[16]   Patel M, Yanagishita M, Roderiquez G, *et al.* Cell-surface heparan sulfate proteoglycan mediates
        HIV-1 infection of T-cell lines. AIDS Res Hum Retroviruses 1993; 9(2): 167-74.
        [http://dx.doi.org/10.1089/aid.1993.9.167] [PMID: 8096145]

[17]   Chen Y, Maguire T, Hileman RE, *et al.* Dengue virus infectivity depends on envelope protein binding
        to target cell heparan sulfate. Nat Med 1997; 3(8): 866-71.
        [http://dx.doi.org/10.1038/nm0897-866] [PMID: 9256277]

[18]   Salvador B, Sexton NR, Carrion R Jr, *et al.* Filoviruses utilize glycosaminoglycans for their
        attachment to target cells. J Virol 2013; 87(6): 3295-304.
        [http://dx.doi.org/10.1128/JVI.01621-12] [PMID: 23302881]

[19]   Byrnes AP, Griffin DE. Binding of Sindbis virus to cell surface heparan sulfate. J Virol 1998; 72(9):
        7349-56.

[PMID: 9696831]

[20] Krusat T, Streckert HJ. Heparin-dependent attachment of respiratory syncytial virus (RSV) to host cells. Arch Virol 1997; 142(6): 1247-54.
[http://dx.doi.org/10.1007/s007050050156] [PMID: 9229012]

[21] Martínez I, Melero JA. Binding of human respiratory syncytial virus to cells: implication of sulfated cell surface proteoglycans. J Gen Virol 2000; 81(Pt 11): 2715-22.
[http://dx.doi.org/10.1099/0022-1317-81-11-2715] [PMID: 11038384]

[22] Feldman SA, Hendry RM, Beeler JA. Identification of a linear heparin binding domain for human respiratory syncytial virus attachment glycoprotein G. J Virol 1999; 73(8): 6610-7.
[PMID: 10400758]

[23] Mathieu C, Dhondt KP, Châlons M, *et al.* Heparan sulfate-dependent enhancement of henipavirus infection. MBio 2015; 6(2): e02427.
[http://dx.doi.org/10.1128/mBio.02427-14] [PMID: 25759505]

[24] Bose S, Banerjee AK. Role of heparan sulfate in human parainfluenza virus type 3 infection. Virology 2002; 298(1): 73-83.
[http://dx.doi.org/10.1006/viro.2002.1484] [PMID: 12093175]

[25] de Boer SM, Kortekaas J, de Haan CA, Rottier PJ, Moormann RJ, Bosch BJ. Heparan sulfate facilitates Rift Valley fever virus entry into the cell. J Virol 2012; 86(24): 13767-71.
[http://dx.doi.org/10.1128/JVI.01364-12] [PMID: 23015725]

[26] Vicenzi E, Canducci F, Pinna D, *et al.* Coronaviridae and SARS-associated coronavirus strain HSR1. Emerg Infect Dis 2004; 10(3): 413-8.
[http://dx.doi.org/10.3201/eid1003.030683] [PMID: 15109406]

[27] WuDunn D, Spear PG. Initial interaction of herpes simplex virus with cells is binding to heparan sulfate. J Virol 1989; 63(1): 52-8.
[PMID: 2535752]

[28] Herold BC, Visalli RJ, Susmarski N, Brandt CR, Spear PG. Glycoprotein C-independent binding of herpes simplex virus to cells requires cell surface heparan sulphate and glycoprotein B. J Gen Virol 1994; 75(Pt 6): 1211-22.
[http://dx.doi.org/10.1099/0022-1317-75-6-1211] [PMID: 8207388]

[29] Shieh MT, WuDunn D, Montgomery RI, Esko JD, Spear PG. Cell surface receptors for herpes simplex virus are heparan sulfate proteoglycans. J Cell Biol 1992; 116(5): 1273-81.
[http://dx.doi.org/10.1083/jcb.116.5.1273] [PMID: 1310996]

[30] Cheshenko N, Herold BC. Glycoprotein B plays a predominant role in mediating herpes simplex virus type 2 attachment and is required for entry and cell-to-cell spread. J Gen Virol 2002; 83(Pt 9): 2247-55.
[http://dx.doi.org/10.1099/0022-1317-83-9-2247] [PMID: 12185280]

[31] Compton T, Nowlin DM, Cooper NR. Initiation of human cytomegalovirus infection requires initial interaction with cell surface heparan sulfate. Virology 1993; 193(2): 834-41.
[http://dx.doi.org/10.1006/viro.1993.1192] [PMID: 8384757]

[32] Zhu Z, Gershon MD, Ambron R, Gabel C, Gershon AA. Infection of cells by varicella zoster virus:

inhibition of viral entry by mannose 6-phosphate and heparin. Proc Natl Acad Sci USA 1995; 92(8): 3546-50.
[http://dx.doi.org/10.1073/pnas.92.8.3546] [PMID: 7724595]

[33]   Birkmann A, Mahr K, Ensser A, *et al.* Cell surface heparan sulfate is a receptor for human herpesvirus 8 and interacts with envelope glycoprotein K8.1. J Virol 2001; 75(23): 11583-93.
[http://dx.doi.org/10.1128/JVI.75.23.11583-11593.2001] [PMID: 11689640]

[34]   Akula SM, Wang FZ, Vieira J, Chandran B. Human herpesvirus 8 interaction with target cells involves heparan sulfate. Virology 2001; 282(2): 245-55.
[http://dx.doi.org/10.1006/viro.2000.0851] [PMID: 11289807]

[35]   Ho Y, Hsiao JC, Yang MH, *et al.* The oligomeric structure of vaccinia viral envelope protein A27L is essential for binding to heparin and heparan sulfates on cell surfaces: a structural and functional approach using site-specific mutagenesis. J Mol Biol 2005; 349(5): 1060-71.
[http://dx.doi.org/10.1016/j.jmb.2005.04.024] [PMID: 15913650]

[36]   Dechecchi MC, Tamanini A, Bonizzato A, Cabrini G. Heparan sulfate glycosaminoglycans are involved in adenovirus type 5 and 2-host cell interactions. Virology 2000; 268(2): 382-90.
[http://dx.doi.org/10.1006/viro.1999.0171] [PMID: 10704346]

[37]   Dechecchi MC, Melotti P, Bonizzato A, Santacatterina M, Chilosi M, Cabrini G. Heparan sulfate glycosaminoglycans are receptors sufficient to mediate the initial binding of adenovirus types 2 and 5. J Virol 2001; 75(18): 8772-80.
[http://dx.doi.org/10.1128/JVI.75.18.8772-8780.2001] [PMID: 11507222]

[38]   Tamura M, Natori K, Kobayashi M, Miyamura T, Takeda N. Genogroup II noroviruses efficiently bind to heparan sulfate proteoglycan associated with the cellular membrane. J Virol 2004; 78(8): 3817-26.
[http://dx.doi.org/10.1128/JVI.78.8.3817-3826.2004] [PMID: 15047797]

[39]   Giroglou T, Florin L, Schäfer F, Streeck RE, Sapp M. Human papillomavirus infection requires cell surface heparan sulfate. J Virol 2001; 75(3): 1565-70.
[http://dx.doi.org/10.1128/JVI.75.3.1565-1570.2001] [PMID: 11152531]

[40]   Schowalter RM, Pastrana DV, Buck CB. Glycosaminoglycans and sialylated glycans sequentially facilitate Merkel cell polyomavirus infectious entry. PLoS Pathog 2011; 7(7): e1002161.
[http://dx.doi.org/10.1371/journal.ppat.1002161] [PMID: 21829355]

[41]   Conti C, Mastromarino P, Riccioli A, Orsi N. Electrostatic interactions in the early events of VSV infection. Res Virol 1991; 142(1): 17-24.
[http://dx.doi.org/10.1016/0923-2516(91)90023-V] [PMID: 1647050]

[42]   Shukla D, Spear PG. Herpesviruses and heparan sulfate: an intimate relationship in aid of viral entry. J Clin Invest 2001; 108(4): 503-10.
[http://dx.doi.org/10.1172/JCI200113799] [PMID: 11518721]

[43]   Herold BC, WuDunn D, Soltys N, Spear PG. Glycoprotein C of herpes simplex virus type 1 plays a principal role in the adsorption of virus to cells and in infectivity. J Virol 1991; 65(3): 1090-8.
[PMID: 1847438]

[44]   Laquerre S, Argnani R, Anderson DB, Zucchini S, Manservigi R, Glorioso JC. Heparan sulfate

proteoglycan binding by herpes simplex virus type 1 glycoproteins B and C, which differ in their contributions to virus attachment, penetration, and cell-to-cell spread. J Virol 1998; 72(7): 6119-30. [PMID: 9621076]

[45]   Trybala E, Bergström T, Svennerholm B, Jeansson S, Glorioso JC, Olofsson S. Localization of a functional site on herpes simplex virus type 1 glycoprotein C involved in binding to cell surface heparan sulphate. J Gen Virol 1994; 75(Pt 4): 743-52. [http://dx.doi.org/10.1099/0022-1317-75-4-743] [PMID: 7512117]

[46]   Mårdberg K, Trybala E, Glorioso JC, Bergström T. Mutational analysis of the major heparan sulfate-binding domain of herpes simplex virus type 1 glycoprotein C. J Gen Virol 2001; 82(Pt 8): 1941-50. [http://dx.doi.org/10.1099/0022-1317-82-8-1941] [PMID: 11458001]

[47]   Penin F, Combet C, Germanidis G, Frainais PO, Deléage G, Pawlotsky JM. Conservation of the conformation and positive charges of hepatitis C virus E2 envelope glycoprotein hypervariable region 1 points to a role in cell attachment. J Virol 2001; 75(12): 5703-10. [http://dx.doi.org/10.1128/JVI.75.12.5703-5710.2001] [PMID: 11356980]

[48]   Crublet E, Andrieu JP, Vivès RR, Lortat-Jacob H. The HIV-1 envelope glycoprotein gp120 features four heparan sulfate binding domains, including the co-receptor binding site. J Biol Chem 2008; 283(22): 15193-200. [http://dx.doi.org/10.1074/jbc.M800066200] [PMID: 18378683]

[49]   Vivès RR, Imberty A, Sattentau QJ, Lortat-Jacob H. Heparan sulfate targets the HIV-1 envelope glycoprotein gp120 coreceptor binding site. J Biol Chem 2005; 280(22): 21353-7. [http://dx.doi.org/10.1074/jbc.M500911200] [PMID: 15797855]

[50]   Shih PC, Yang MS, Lin SC, *et al.* A turn-like structure KKPE segment mediates the specific binding of viral protein A27 to heparin and heparan sulfate on cell surfaces. J Biol Chem 2009; 284(52): 36535-46. [http://dx.doi.org/10.1074/jbc.M109.037267] [PMID: 19858217]

[51]   Joyce JG, Tung JS, Przysiecki CT, *et al.* The L1 major capsid protein of human papillomavirus type 11 recombinant virus-like particles interacts with heparin and cell-surface glycosaminoglycans on human keratinocytes. J Biol Chem 1999; 274(9): 5810-22. [http://dx.doi.org/10.1074/jbc.274.9.5810] [PMID: 10026203]

[52]   Hirst GK. The agglutination of red cells by allantoic fluid of chick embryos infected with influenza virus. Science 1941; 94(2427): 22-3. [http://dx.doi.org/10.1126/science.94.2427.22] [PMID: 17777315]

[53]   Haff RF, Stewart RC. Role of sialic acid receptors in adsorption of influenza virus to chick embryo cells. J Immunol 1965; 94: 842-51. [PMID: 14321427]

[54]   Wang Q, Tian X, Chen X, Ma J. Structural basis for receptor specificity of influenza B virus hemagglutinin. Proc Natl Acad Sci USA 2007; 104(43): 16874-9. [http://dx.doi.org/10.1073/pnas.0708363104] [PMID: 17942670]

[55]   Rogers GN, Herrler G, Paulson JC, Klenk HD. Influenza C virus uses 9-O-acetyl-N-acetylneuraminic acid as a high affinity receptor determinant for attachment to cells. J Biol Chem 1986; 261(13): 5947-51.

[PMID: 3700379]

[56]   Vlasak R, Luytjes W, Spaan W, Palese P. Human and bovine coronaviruses recognize sialic acid-containing receptors similar to those of influenza C viruses. Proc Natl Acad Sci USA 1988; 85(12): 4526-9.
[http://dx.doi.org/10.1073/pnas.85.12.4526] [PMID: 3380803]

[57]   Suzuki Y, Hirabayashi Y, Suzuki T, Matsumoto M. Occurrence of O-glycosidically peptide-linked oligosaccharides of poly-N-acetyllactosamine type (erythroglycan II) in the I-antigenically active Sendai virus receptor sialoglycoprotein GP-2. J Biochem 1985; 98(6): 1653-9.
[PMID: 3005249]

[58]   Rydell GE, Nilsson J, Rodriguez-Diaz J, *et al.* Human noroviruses recognize sialyl Lewis x neoglycoprotein. Glycobiology 2009; 19(3): 309-20.
[http://dx.doi.org/10.1093/glycob/cwn139] [PMID: 19054801]

[59]   Gentsch JR, Pacitti AF. Effect of neuraminidase treatment of cells and effect of soluble glycoproteins on type 3 reovirus attachment to murine L cells. J Virol 1985; 56(2): 356-64.
[PMID: 4057353]

[60]   Nokhbeh MR, Hazra S, Alexander DA, *et al.* Enterovirus 70 binds to different glycoconjugates containing alpha2,3-linked sialic acid on different cell lines. J Virol 2005; 79(11): 7087-94.
[http://dx.doi.org/10.1128/JVI.79.11.7087-7094.2005] [PMID: 15890948]

[61]   Liu CK, Wei G, Atwood WJ. Infection of glial cells by the human polyomavirus JC is mediated by an N-linked glycoprotein containing terminal alpha(26)-linked sialic acids. J Virol 1998; 72(6): 4643-9.
[PMID: 9573227]

[62]   Dugan AS, Eash S, Atwood WJ. An N-linked glycoprotein with alpha(2,3)-linked sialic acid is a receptor for BK virus. J Virol 2005; 79(22): 14442-5.
[http://dx.doi.org/10.1128/JVI.79.22.14442-14445.2005] [PMID: 16254379]

[63]   Lehmann F, Tiralongo E, Tiralongo J. Sialic acid-specific lectins: occurrence, specificity and function. Cell Mol Life Sci 2006; 63(12): 1331-54.
[http://dx.doi.org/10.1007/s00018-005-5589-y] [PMID: 16596337]

[64]   Uncapher CR, DeWitt CM, Colonno RJ. The major and minor group receptor families contain all but one human rhinovirus serotype. Virology 1991; 180(2): 814-7.
[http://dx.doi.org/10.1016/0042-6822(91)90098-V] [PMID: 1846502]

[65]   Nilsson EC, Jamshidi F, Johansson SM, Oberste MS, Arnberg N. Sialic acid is a cellular receptor for coxsackievirus A24 variant, an emerging virus with pandemic potential. J Virol 2008; 82(6): 3061-8.
[http://dx.doi.org/10.1128/JVI.02470-07] [PMID: 18184708]

[66]   Arnberg N, Edlund K, Kidd AH, Wadell G. Adenovirus type 37 uses sialic acid as a cellular receptor. J Virol 2000; 74(1): 42-8.
[http://dx.doi.org/10.1128/JVI.74.1.42-48.2000] [PMID: 10590089]

[67]   Arnberg N, Pring-Akerblom P, Wadell G. Adenovirus type 37 uses sialic acid as a cellular receptor on Chang C cells. J Virol 2002; 76(17): 8834-41.
[http://dx.doi.org/10.1128/JVI.76.17.8834-8841.2002] [PMID: 12163603]

[68]   Weis W, Brown JH, Cusack S, Paulson JC, Skehel JJ, Wiley DC. Structure of the influenza virus

haemagglutinin complexed with its receptor, sialic acid. Nature 1988; 333(6172): 426-31.
[http://dx.doi.org/10.1038/333426a0] [PMID: 3374584]

[69]   Connaris H, Takimoto T, Russell R, *et al.* Probing the sialic acid binding site of the hemagglutinin-neuraminidase of Newcastle disease virus: identification of key amino acids involved in cell binding, catalysis, and fusion. J Virol 2002; 76(4): 1816-24.
       [http://dx.doi.org/10.1128/JVI.76.4.1816-1824.2002] [PMID: 11799177]

[70]   Reiter DM, Frierson JM, Halvorson EE, Kobayashi T, Dermody TS, Stehle T. Crystal structure of reovirus attachment protein σ1 in complex with sialylated oligosaccharides. PLoS Pathog 2011; 7(8): e1002166.
       [http://dx.doi.org/10.1371/journal.ppat.1002166] [PMID: 21829363]

[71]   Matrosovich MN. Towards the development of antimicrobial drugs acting by inhibition of pathogen attachment to host cells: a need for polyvalency. FEBS Lett 1989; 252(1-2): 1-4.
       [http://dx.doi.org/10.1016/0014-5793(89)80879-8] [PMID: 2668026]

[72]   Baba M, Snoeck R, Pauwels R, de Clercq E. Sulfated polysaccharides are potent and selective inhibitors of various enveloped viruses, including herpes simplex virus, cytomegalovirus, vesicular stomatitis virus, and human immunodeficiency virus. Antimicrob Agents Chemother 1988; 32(11): 1742-5.
       [http://dx.doi.org/10.1128/AAC.32.11.1742] [PMID: 2472775]

[73]   Lee E, Pavy M, Young N, Freeman C, Lobigs M. Antiviral effect of the heparan sulfate mimetic, PI-88, against dengue and encephalitic flaviviruses. Antiviral Res 2006; 69(1): 31-8.
       [http://dx.doi.org/10.1016/j.antiviral.2005.08.006] [PMID: 16309754]

[74]   Clanton DJ, Moran RA, McMahon JB, *et al.* Sulfonic acid dyes: inhibition of the human immunodeficiency virus and mechanism of action. J Acquir Immune Defic Syndr 1992; 5(8): 771-81.
       [PMID: 1517963]

[75]   Mohan P, Schols D, Baba M, De Clercq E. Sulfonic acid polymers as a new class of human immunodeficiency virus inhibitors. Antiviral Res 1992; 18(2): 139-50.
       [http://dx.doi.org/10.1016/0166-3542(92)90034-3] [PMID: 1384428]

[76]   Aguilar JS, Rice M, Wagner EK. The polysulfonated compound suramin blocks adsorption and lateral diffusion of herpes simplex virus type-1 in vero cells. Virology 1999; 258(1): 141-51.
       [http://dx.doi.org/10.1006/viro.1999.9723] [PMID: 10329576]

[77]   Basu A, Kanda T, Beyene A, Saito K, Meyer K, Ray R. Sulfated homologues of heparin inhibit hepatitis C virus entry into mammalian cells. J Virol 2007; 81(8): 3933-41.
       [http://dx.doi.org/10.1128/JVI.02622-06] [PMID: 17287282]

[78]   Moulard M, Lortat-Jacob H, Mondor I, *et al.* Selective interactions of polyanions with basic surfaces on human immunodeficiency virus type 1 gp120. J Virol 2000; 74(4): 1948-60.
       [http://dx.doi.org/10.1128/JVI.74.4.1948-1960.2000] [PMID: 10644368]

[79]   Schols D, Baba M, Pauwels R, Desmyter J, De Clercq E. Specific interaction of aurintricarboxylic acid with the human immunodeficiency virus/CD4 cell receptor. Proc Natl Acad Sci USA 1989; 86(9): 3322-6.
       [http://dx.doi.org/10.1073/pnas.86.9.3322] [PMID: 2566170]

[80]   Schols D, Wutzler P, Klöcking R, Helbig B, De Clercq E. Selective inhibitory activity of polyhydroxycarboxylates derived from phenolic compounds against human immunodeficiency virus replication. J Acquir Immune Defic Syndr 1991; 4(7): 677-85.
[PMID: 1675677]

[81]   Meerbach A, Neyts J, Balzarini J, Helbig B, De Clercq E, Wutzler P. *In vitro* activity of polyhydroxycarboxylates against herpesviruses and HIV. Antivir Chem Chemother 2001; 12(6): 337-45.
[http://dx.doi.org/10.1177/095632020101200603] [PMID: 12018678]

[82]   Helbig BK, Wutzler P. Anti-herpes simplex virus type 1 activity of humic acid-like polymers and their o-diphenolic starting compounds. Antivir Chem Chemother 1997; 8: 265-73.
[http://dx.doi.org/10.1177/095632029700800310]

[83]   Krepstakies M, Lucifora J, Nagel CH, *et al.* A new class of synthetic peptide inhibitors blocks attachment and entry of human pathogenic viruses. J Infect Dis 2012; 205(11): 1654-64.
[http://dx.doi.org/10.1093/infdis/jis273] [PMID: 22457281]

[84]   Lorentsen KJ, Hendrix CW, Collins JM, *et al.* Dextran sulfate is poorly absorbed after oral administration. Ann Intern Med 1989; 111(7): 561-6.
[http://dx.doi.org/10.7326/0003-4819-111-7-561] [PMID: 2476054]

[85]   Flexner C, Barditch-Crovo PA, Kornhauser DM, *et al.* Pharmacokinetics, toxicity, and activity of intravenous dextran sulfate in human immunodeficiency virus infection. Antimicrob Agents Chemother 1991; 35(12): 2544-50.
[http://dx.doi.org/10.1128/AAC.35.12.2544] [PMID: 1810188]

[86]   Rosenberg RD. Heparin, antithrombin, and abnormal clotting. Annu Rev Med 1978; 29: 367-78.
[http://dx.doi.org/10.1146/annurev.me.29.020178.002055] [PMID: 77142]

[87]   Baba M, De Clercq E, Schols D, *et al.* Novel sulfated polysaccharides: dissociation of anti-human immunodeficiency virus activity from antithrombin activity. J Infect Dis 1990; 161(2): 208-13.
[http://dx.doi.org/10.1093/infdis/161.2.208] [PMID: 2405068]

[88]   Bârzu T, Level M, Petitou M, *et al.* Preparation and anti-HIV activity of O-acylated heparin and dermatan sulfate derivatives with low anticoagulant effect. J Med Chem 1993; 36(23): 3546-55.
[http://dx.doi.org/10.1021/jm00075a009] [PMID: 8246223]

[89]   Witvrouw M, De Clercq E. Sulfated polysaccharides extracted from sea algae as potential antiviral drugs. Gen Pharmacol 1997; 29(4): 497-511.
[http://dx.doi.org/10.1016/S0306-3623(96)00563-0] [PMID: 9352294]

[90]   Harden EA, Falshaw R, Carnachan SM, Kern ER, Prichard MN. Virucidal activity of polysaccharide extracts from four algal species against herpes simplex virus. Antiviral Res 2009; 83(3): 282-9.
[http://dx.doi.org/10.1016/j.antiviral.2009.06.007] [PMID: 19576248]

[91]   Lin LT, Chen TY, Lin SC, *et al.* Broad-spectrum antiviral activity of chebulagic acid and punicalagin against viruses that use glycosaminoglycans for entry. BMC Microbiol 2013; 13: 187.
[http://dx.doi.org/10.1186/1471-2180-13-187] [PMID: 23924316]

[92]   Lin LT, Chen TY, Chung CY, *et al.* Hydrolyzable tannins (chebulagic acid and punicalagin) target viral glycoprotein-glycosaminoglycan interactions to inhibit herpes simplex virus 1 entry and cell-t-

-cell spread. J Virol 2011; 85(9): 4386-98.
[http://dx.doi.org/10.1128/JVI.01492-10] [PMID: 21307190]

[93] Matrosovich M, Klenk HD. Natural and synthetic sialic acid-containing inhibitors of influenza virus receptor binding. Rev Med Virol 2003; 13(2): 85-97.
[http://dx.doi.org/10.1002/rmv.372] [PMID: 12627392]

[94] Pritchett TJ, Paulson JC. Basis for the potent inhibition of influenza virus infection by equine and guinea pig alpha 2-macroglobulin. J Biol Chem 1989; 264(17): 9850-8.
[PMID: 2470765]

[95] Mammen M, Dahmann G, Whitesides GM. Effective inhibitors of hemagglutination by influenza virus synthesized from polymers having active ester groups. Insight into mechanism of inhibition. J Med Chem 1995; 38(21): 4179-90.
[http://dx.doi.org/10.1021/jm00021a007] [PMID: 7473545]

[96] Lees WJ, Spaltenstein A, Kingery-Wood JE, Whitesides GM. Polyacrylamides bearing pendant alpha-sialoside groups strongly inhibit agglutination of erythrocytes by influenza A virus: multivalency and steric stabilization of particulate biological systems. J Med Chem 1994; 37(20): 3419-33.
[http://dx.doi.org/10.1021/jm00046a027] [PMID: 7932570]

[97] Reuter JD, Myc A, Hayes MM, et al. Inhibition of viral adhesion and infection by sialic-aci--conjugated dendritic polymers. Bioconjug Chem 1999; 10(2): 271-8.
[http://dx.doi.org/10.1021/bc980099n] [PMID: 10077477]

[98] Hidari KI, Murata T, Yoshida K, et al. Chemoenzymatic synthesis, characterization, and application of glycopolymers carrying lactosamine repeats as entry inhibitors against influenza virus infection. Glycobiology 2008; 18(10): 779-88.
[http://dx.doi.org/10.1093/glycob/cwn067] [PMID: 18621993]

[99] Waldmann M, Jirmann R, Hoelscher K, et al. A nanomolar multivalent ligand as entry inhibitor of the hemagglutinin of avian influenza. J Am Chem Soc 2014; 136(2): 783-8.
[http://dx.doi.org/10.1021/ja410918a] [PMID: 24377426]

[100] Sparks MA, Williams KW, Whitesides GM. Neuraminidase-resistant hemagglutination inhibitors: acrylamide copolymers containing a C-glycoside of N-acetylneuraminic acid. J Med Chem 1993; 36(6): 778-83.
[http://dx.doi.org/10.1021/jm00058a016] [PMID: 8459405]

[101] Itoh M, Hetterich P, Isecke R, Brossmer R, Klenk HD. Suppression of influenza virus infection by an N-thioacetylneuraminic acid acrylamide copolymer resistant to neuraminidase. Virology 1995; 212(2): 340-7.
[http://dx.doi.org/10.1006/viro.1995.1491] [PMID: 7571403]

[102] Guo CT, Sun XL, Kanie O, et al. An O-glycoside of sialic acid derivative that inhibits both hemagglutinin and sialidase activities of influenza viruses. Glycobiology 2002; 12(3): 183-90.
[http://dx.doi.org/10.1093/glycob/12.3.183] [PMID: 11971862]

[103] Moss RB, Hansen C, Sanders RL, Hawley S, Li T, Steigbigel RT. A phase II study of DAS181, a novel host directed antiviral for the treatment of influenza infection. J Infect Dis 2012; 206(12): 1844-51.
[http://dx.doi.org/10.1093/infdis/jis622] [PMID: 23045618]

[104] Malakhov MP, Aschenbrenner LM, Smee DF, *et al.* Sialidase fusion protein as a novel broad-spectrum inhibitor of influenza virus infection. Antimicrob Agents Chemother 2006; 50(4): 1470-9.
[http://dx.doi.org/10.1128/AAC.50.4.1470-1479.2006] [PMID: 16569867]

[105] Belser JA, Lu X, Szretter KJ, *et al.* DAS181, a novel sialidase fusion protein, protects mice from lethal avian influenza H5N1 virus infection. J Infect Dis 2007; 196(10): 1493-9.
[http://dx.doi.org/10.1086/522609] [PMID: 18008229]

[106] Nicholls JM, Moss RB, Haslam SM. The use of sialidase therapy for respiratory viral infections. Antiviral Res 2013; 98(3): 401-9.
[http://dx.doi.org/10.1016/j.antiviral.2013.04.012] [PMID: 23602850]

[107] Steinmann J, Buer J, Pietschmann T, Steinmann E. Anti-infective properties of epigallocatechin--gallate (EGCG), a component of green tea. Br J Pharmacol 2013; 168(5): 1059-73.
[http://dx.doi.org/10.1111/bph.12009] [PMID: 23072320]

[108] Nance CL, Siwak EB, Shearer WT. Preclinical development of the green tea catechin, epigallocatechin gallate, as an HIV-1 therapy. J Allergy Clin Immunol 2009; 123(2): 459-65.
[http://dx.doi.org/10.1016/j.jaci.2008.12.024] [PMID: 19203663]

[109] Yamaguchi K, Honda M, Ikigai H, Hara Y, Shimamura T. Inhibitory effects of (-)-epigallocatechin gallate on the life cycle of human immunodeficiency virus type 1 (HIV-1). Antiviral Res 2002; 53(1): 19-34.
[http://dx.doi.org/10.1016/S0166-3542(01)00189-9] [PMID: 11684313]

[110] Weber JM, Ruzindana-Umunyana A, Imbeault L, Sircar S. Inhibition of adenovirus infection and adenain by green tea catechins. Antiviral Res 2003; 58(2): 167-73.
[http://dx.doi.org/10.1016/S0166-3542(02)00212-7] [PMID: 12742577]

[111] Song JM, Lee KH, Seong BL. Antiviral effect of catechins in green tea on influenza virus. Antiviral Res 2005; 68(2): 66-74.
[http://dx.doi.org/10.1016/j.antiviral.2005.06.010] [PMID: 16137775]

[112] Xu J, Wang J, Deng F, Hu Z, Wang H. Green tea extract and its major component epigallocatechin gallate inhibits hepatitis B virus *in vitro.* Antiviral Res 2008; 78(3): 242-9.
[http://dx.doi.org/10.1016/j.antiviral.2007.11.011] [PMID: 18313149]

[113] Huang HC, Tao MH, Hung TM, Chen JC, Lin ZJ, Huang C. (-)-Epigallocatechin-3-gallate inhibits entry of hepatitis B virus into hepatocytes. Antiviral Res 2014; 111: 100-11.
[http://dx.doi.org/10.1016/j.antiviral.2014.09.009] [PMID: 25260897]

[114] Ciesek S, von Hahn T, Colpitts CC, *et al.* The green tea polyphenol, epigallocatechin-3-gallate, inhibits hepatitis C virus entry. Hepatology 2011; 54(6): 1947-55.
[http://dx.doi.org/10.1002/hep.24610] [PMID: 21837753]

[115] Isaacs CE, Wen GY, Xu W, *et al.* Epigallocatechin gallate inactivates clinical isolates of herpes simplex virus. Antimicrob Agents Chemother 2008; 52(3): 962-70.
[http://dx.doi.org/10.1128/AAC.00825-07] [PMID: 18195068]

[116] Isaacs CE, Xu W, Merz G, Hillier S, Rohan L, Wen GY. Digallate dimers of (-)-epigallocatechin gallate inactivate herpes simplex virus. Antimicrob Agents Chemother 2011; 55(12): 5646-53.
[http://dx.doi.org/10.1128/AAC.05531-11] [PMID: 21947401]

[117]   Kim M, Kim SY, Lee HW, *et al.* Inhibition of influenza virus internalization by (-)-epigallocatechi-
        -3-gallate. Antiviral Res 2013; 100(2): 460-72.
        [http://dx.doi.org/10.1016/j.antiviral.2013.08.002] [PMID: 23954192]

[118]   Colpitts CC, Schang LM. A small molecule inhibits virion attachment to heparan sulfate- or sialic
        acid-containing glycans. J Virol 2014; 88(14): 7806-17.
        [http://dx.doi.org/10.1128/JVI.00896-14] [PMID: 24789779]

[119]   Ge H, Liu G, Xiang YF, *et al.* The mechanism of poly-galloyl-glucoses preventing Influenza A virus
        entry into host cells. PLoS One 2014; 9(4): e94392.
        [http://dx.doi.org/10.1371/journal.pone.0094392] [PMID: 24718639]

[120]   Liu G, Xiong S, Xiang YF, *et al.* Antiviral activity and possible mechanisms of action of
        pentagalloylglucose (PGG) against influenza A virus. Arch Virol 2011; 156(8): 1359-69.
        [http://dx.doi.org/10.1007/s00705-011-0989-9] [PMID: 21479599]

[121]   Lee SJ, Lee HK, Jung MK, Mar W. *In vitro* antiviral activity of 1,2,3,4,6-penta-O-galloyl-b-
        ta-D-glucose against hepatitis B virus. Biol Pharm Bull 2006; 29(10): 2131-4.
        [http://dx.doi.org/10.1248/bpb.29.2131] [PMID: 17015965]

[122]   Smith TJ. Green Tea Polyphenols in drug discovery - a success or failure? Expert Opin Drug Discov
        2011; 6(6): 589-95.
        [http://dx.doi.org/10.1517/17460441.2011.570750] [PMID: 21731575]

CHAPTER 3

Carbohydrate PEGylation in Chemotherapy

Rosalia Agusti, M. Eugenia Giorgi and **Rosa M. de Lederkremer**[*]

Universidad de Buenos Aires, Consejo Nacional de Investigaciones Científicas y Técnicas, Centro de Investigación en Hidratos de Carbono (CIHIDECAR), Facultad de Ciencias Exactas y Naturales, Buenos Aires, Argentina

Abstract: Improvement of drug delivery by covalent modification with polyethylene glycol (PEG) has been widely applied to proteins and low molecular weight pharmaceuticals. Protein therapeutic agents have the drawback of short circulating half-life due to their lability towards proteolytic enzymes and rapid clearance by kidney filtration. Moreover, their recognition by the immune system results in the production of neutralizing antibodies. Several strategies have been used for the PEGylation of proteins at the amino group of lysine but the product is heterogeneous due to random PEGylation. More recently, selective PEGylation of proteins was achieved by introduction of a PEGylated carbohydrate in an *O*-glycosylation site. This technique is called glycoPEGylation and is used for the production of several therapeutics, some of them are currently in advanced clinical trials. Improvement of the delivery system rather than the drug itself has led to the optimization of the therapeutic properties of existing drugs by minimizing their side effects. In this respect, active targeting to receptors present in cancer tissues prevents healthy ones from being damaged. Since carbohydrates have been recognized as playing an important role in the interaction with cell receptors, PEGylated carbohydrates have been used for active targeting of drugs. In this review we discuss different applications of sugar-PEGylation.

Keywords: Active targeting, Bioavailability, Carbohydrates, Chemotherapy, Drug carriers, Drug delivery, Gene delivery, Gene therapy, Glycan delivery, GlycoPEGylation, Ligands, Multiarm PEGs, O-GlcNAc quantification, PEG, PEG-sialic acid, PEGylated nanoparticles, PEGylated polysaccharides,

[*] **Corresponding author Rosa M. de Lederkremer:** CIHIDECAR-CONICET, Departamento de Química Orgánica, Facultad de Ciencias Exactas y Naturales, Universidad de Buenos Aires, Buenos Aires, Argentina; Tel/Fax: 54-1--45763352; E-mail: lederk@qo.fcen.uba.ar

Qun Zhou (Ed.)

PEGylation, siRNA, Sugar PEGylation, Targeted delivery.

INTRODUCTION

Bioavailability of drugs in circulation remains a crucial issue in chemotherapy since most of them are degraded by metabolic processes or are poorly soluble in aqueous systems. The most well-established and widely employed way to increase their circulation half-life without affecting their activity is by conjugation with a water soluble biocompatible polymer. In this regard, poly(ethyleneglycol) (PEG) has been extensively used due to its lack of toxicity and immunogenicity. These characteristics earned PEG its approval by the United States Food and Drug Administration (USFDA) for its use as excipient in pharmaceutical formulas [1].

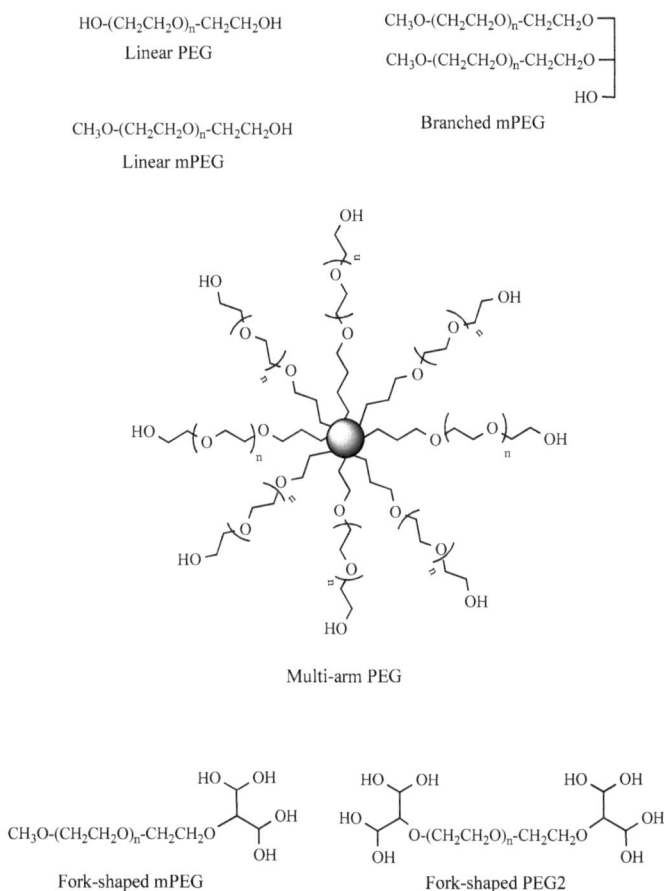

Fig. (1). Types of PEG used for the derivatization of drugs and proteins.

PEG reagents are commercially available in different lengths, shapes and chemistries, allowing them to react with particular functional groups for their covalent attachment. PEGs may be arranged in linear or branched structures, including multiarm PEGs designed to increase the loading of the active molecule (Fig. **1**).

PEG derivatives were first developed for the attachment to the amino groups in proteins [2] and bovine serum albumin was one of the first to be linked to a methoxypolyethylen glycol (mPEG) chain [3]. In this and some other early works the amino groups at the N terminus or in lysine side chains were used for PEG attachment with activated carboxyl or aldehyde functionalized PEGs. Currently, there are many monofunctional and homo and heterobifunctional PEGs available in the market for amide linking including *N*-hydroxysuccinimidyl esters (PEG-NHS), carboxylic acids (PEG-COOH) for *in situ* activation with carbodiimide (EDC or DCC), and *p*-nitrophenylcarbonates (PEG-NPC) (Fig. **2**).

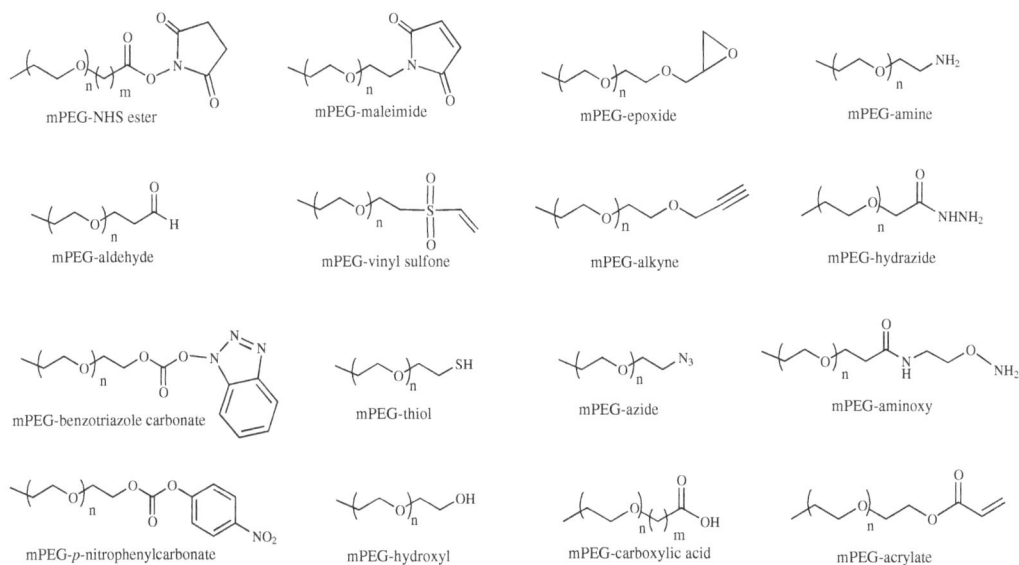

mPEG-NHS ester · mPEG-maleimide · mPEG-epoxide · mPEG-amine
mPEG-aldehyde · mPEG-vinyl sulfone · mPEG-alkyne · mPEG-hydrazide
mPEG-benzotriazole carbonate · mPEG-thiol · mPEG-azide · mPEG-aminoxy
mPEG-*p*-nitrophenylcarbonate · mPEG-hydroxyl · mPEG-carboxylic acid · mPEG-acrylate

Fig. (2). PEG derivatives for conjugation with drugs.

However, the abundance of lysine and its usual surface location lead to complex mixtures differing in the number and site of PEGylation [4]. To avoid these shortcomings, PEGylation has evolved towards site-specific amino acid conjugation approaches that minimize the heterogeneity, ensure batch-to-batch reproducibility and yield well defined products. Examples of such approaches are PEGylation of the amino group at the N-terminus or of the thiol group present in natural or genetically introduced unpaired cysteines.

Selective PEGylation at the N-terminal amino group could be performed by reductive alkylation with an aldehyde functionalized PEG (PEG-CHO) and a reducing agent (*e.g.*, sodium cyanoborohydride), exploiting the different pK_a values of protein amino groups. This strategy has been successfully applied to produce the N-terminally PEGylated recombinant granulocyte colony factor (G-CSF) Filgrastim, marketed under the brand name Neulasta® (pegfilgrastim) [5].

For cysteine modification, a variety of thiol-specific reagents are available, such as maleimide (PEG-MAL), ortho-pyridyl disulfide (PEG-OPSS), vinyl sulfone (PEG-VS), *etc* [6, 7]. Due to the stability of the product, maleimide-PEG reagents have become very popular. A PEG- monoclonal antibody to tumor necrosis factor alpha (TNF-α) (Certolizumab pegol, tradename Cimzia), used for the treatment of Crohn's disease and rheumatoid arthritis, was developed based on this strategy [8].

In addition to chemical procedures, enzymes have also been employed to achieve specific PEGylation. The most common method uses a transglutaminase (TGase, EC 2.3.2.13), capable of catalyzing the incorporation of PEG-amine (PEG-NH$_2$) reagents into the protein natural or genetically introduced glutamine residues [9]. Another monoPEGylated derivative of Filgrastim was produced by using this technique [10]. Since the early studies many PEGylated proteins have been commercialized by pharmaceutical companies, including anti-cancer drugs [11] (Table **1**).

In the last years a method called GlycoPEGylation was conceived for introduction of a PEGylated carbohydrate at *O*-glycosylation sites. Using different strategies, the PEGylated sugar can be introduced by enzymatic methods into glycosylated or non-glycosylated proteins [23]. It will be discussed with more details in the next section.

A carbohydrate may be introduced by chemical ligation to a natural protein followed by periodate oxidation to generate the aldehyde groups suitable for selective PEGylation [24]. Also, the carbohydrate in a native glycoprotein may be enzymatically [25] or chemically [26, 27] modified for further PEGylation.

Table 1. PEG-conjugates approved for therapeutic use.

Market Name	Protein	Year of Approval	Use	PEG Used	Ref.
Adagen®	Adenosine Deaminase	1990	SCID	NHS-ester	[12]
Oncaspar®	L-asparaginase	1994	Leukemia	NHS-ester	[13]
PegIntron®	Interferon α-2b	2000	Hepatitis C	NHS-ester	[14]
Pegasys®	Interferon α-2a	2001	Hepatitis C	NHS-ester	[15]
Neulasta™	G-CSF	2002	Neutropenia	aldehyde	[16]
Somavert™	hGH	2003	Acromegaly	NHS-ester	[17]
Macugen™	Aptamer	2004	Age-related macular degeneration	NHS-ester	[18]
Mircera®	Erythropoietin	2007	Anemia	NHS-ester	[19]
Cimzia®	Anti TNF α Fab'	2008	Rheumatoid arthritis, Crohn disease, psoriatic arthritis	maleimide	[8]
Krystexxa®	Urate oxidase	2010	Gout	*p*-Nitrophenyl carbonate ester	[20]
Omontys®	Synthetic, dimeric peptide (erythropoiesis stimulating agent)	2012 (recall 2014)	Anemia in chronic kidney disease	NHS-ester	[21]
Lonquex®	G-CSF	2013	Neutropenia	*p*-Nitrophenyl carbonate ester	[22]

Although the primary use of PEGylation has been to improve the physicochemical properties of large molecules, it may also be used for small drugs. Indeed, some of the conjugates have entered the clinical trials. PEG-camptothecin (PRO-THECAN®) is in phase 2 clinical trials. The prodrug conjugate was obtained by coupling two molecules of camptothecin to a glycine-bifunctionalised 40 kDa PEG, doubling the loading capacity and increasing the drug half-life in blood [28]. Paclitaxel (taxol ®), a chemotherapy drug used to treat different cancers including ovarian, breast and non-small cell lung cancer, was PEGylated and the product is currently in phase 1 trials [29]. PEG-Intron is used to eradicate hepatic and extrahepatic hepatitis C virus infection. PEG conjugated with α-interferon (IFN) was approved by USFDA for use in 2001 [30].

Most drug or gene delivery systems are comprised of a carrier system conjugated to a drug or gene through a linker. Among others, recombinant viruses are being used as vectors for gene therapy or vaccination. However, its use is hindered by the immune response to the proteins in the virus coat. The development of PEGylated viruses, with attenuated immunogenicity might propel its clinical use. In this respect, Eto and coworkers observed *in vivo* reduction of antibodies against adenovirus after injection of the PEGylated virus [31].

In the last decades the role of carbohydrates in cell recognition and thus in tissue targeting has been frequently addressed. For instance, the asialoglycoprotein receptor for galactose in hepatocytes is a target for therapeutic agents for liver diseases [32, 33]. Also, glycans in the surface of parasites, like the trypanosomatids play an important function in the infection of mammals and the enzymes that recognize them are a target for chemotherapy [34].

The approach of active targeting a sugar receptor led to labeling a drug carrier with a sugar which promotes selective internalization of the drug. Further degradation releases the drug into the cell [35].

A more recent application was developed for quantification of *O*-linked *N*-acetyl glucosamine (*O*-GlcNAc) modification of proteins. This is a post-translational modification found in intracellular proteins which regulates numerous biological processes [36]. The method is based on selectively labeling the sugar with a PEG of defined mass, resolving the labeled populations by one-dimensional sodium dodecyl sulfate-polyacrylamide gel electrophoresis (SDS-PAGE) and identifying them by mass spectrometry [37].

In this review we present examples of modification of glycans by covalent conjugation with PEG reagents used mainly for improvement of the bioavailability of drugs.

SUGAR-PEGYLATION OF PROTEINS

GlycoPEGylation

Proteins offer several advantages over small molecules as therapeutics or as diagnostic probes including exquisite target specificity, multiplicity of function and relatively low off-target activity. The chemical modification of proteins may extend these properties by improving their potency and stability. One significant obstacle to the creation of a chemically altered protein is its production in a biologically active form [38].

Natural sites of glycosylation have been seen as the ideal for PEGylation. A method for conjugation, named GlycoPEGylation, was used to mimic *O*-glycosylation. The strategy is based on the finding that certain PEGylated nucleotide-sugars are effectively transferred to a glycan acceptor by the corresponding glycosyltransferase. According to the Neose Technology procedure [39], in the first step, the hydroxyl group of specific serine or threonine is conjugated to *N*-acetylgalactosamine (GalNAc) by *in vitro* treatment of the recombinant polypeptide with *O*-GalNAc-transferase and UDP-GalNAc. This unit may be further galactosylated by a galactosyltransferase followed by incubation of the galactosylated or the non-galactosylated peptide with the PEGylated sialyl nucleotide (CMP-SA-5-NHCOCH$_2$NHPEG). A chemoenzymatic method for the preparation of the PEGylated nucleotide and incorporation of the PEG-sialic acid in the protein is shown in Scheme **1** [39, 40]. This technique was applied to develop Lipegfilgrastim, the first glycoPEGylated drug approved for therapeutic use in Europe, marketed as Lonquex®. It is a covalent conjugate of recombinant human *N*-methoxy granulocyte-colony stimulating factor (G-CSF), containing a single 20 kDa mPEG attached through two carbohydrate residues to threonine[134] of the peptide backbone [22]. The standard PEGylation technology previously used for filgrastim yielded a heterogeneous product (Pegfilgrastim) with multiple isoforms that required further purification. Moreover, Lipegfilgrastim showed a longer half-life *in vitro* compared with Pegfilgrastim [41].

GlycoPEGylation is certainly the most specific PEGylation method for proteins available to date [42]. Recently, it has also been used in the development of a treatment for hemophilia since several coagulation factors currently in use are naturally glycosylated. In a modified protocol, the sialic acids from the native glycosylated protein were removed by a sialidase and the unmasked galactose units were then sialylated by the action of a sialyltransferase (ST3GalIII) that transfers SA-PEG from CMP-SA-5-NHCOCH$_2$NHPEG. The number of PEG units added depends on the reaction time. After completion, the unreacted galactose residues should be blocked with sialic acid to avoid hepatic clearance by the asialoglycoprotein receptor. GlycoPEGylated factors VIIa and XI have been developed with this approach [43 - 45].

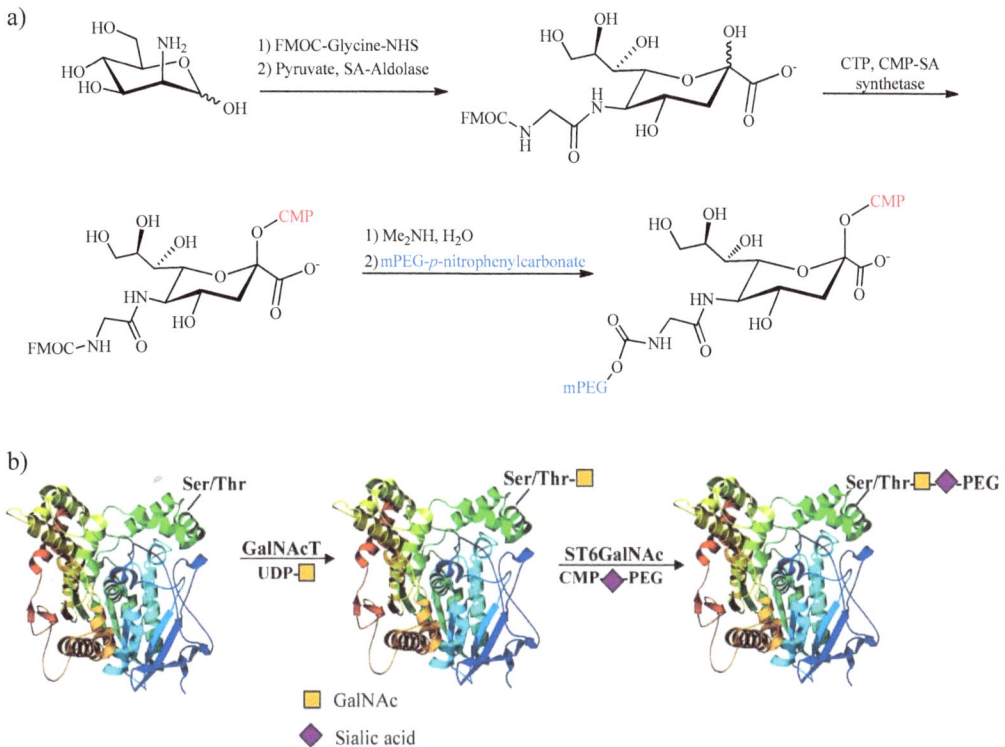

Scheme 1. (a) Synthesis of PEGylated CMP-sialic acid and **(b)** incorporation of PEG-sialic acid into the protein. Adapted from [23].

A variation of this desialylation/resialylation strategy has been applied for PEGylation of a recombinant form of coagulation factor VIII (turoctocog-α), used in the treatment of hemophilia. After removal of all sialic acids, incubation with a specific   sialyltransferase   ST3GalI,   which   selectively   transfers SA-5-NHCOCH$_2$NHPEG to the *O*-glycans, allowed the site-specific modification in the presence of *N*-glycan sites [46].

Others sugar nucleotides, like UDP-Gal-PEG, have been proposed for glycoPEGylation of factor FVII and its activated form FVIIa. In this case, FVII or FVIIa may be first treated with sialidase and galactosidase to produce asialo agalacto polypeptide further PEGylated by incubation with a natural or mutated 1,4- galactosyltransferase [47].

Glycation of Proteins for Further PEGylation

Carbohydrates may be specifically linked to proteins and further conjugated with a PEG chain. Salmaso *et al.* [24] introduced a galactosylglucono sugar moiety through a linker containing a maleimide group in the opposite end. The maleimide covalently binds to the thiol group of Cys[34] in human serum albumin. PEGylation was achieved after periodate oxidation of the sugar residues to create aldehyde groups for multiple coupling with PEG-hydrazide (PEG-Hz) (Scheme **2**). This versatile approach combines the benefits of site selective conjugation at an inaccessible aminoacid of a natural protein with multiple polymer coupling.

PEGylation of Native Glycoproteins

PEGylation of native glycoproteins may be performed by enzymatic or chemical modification of the glycan. A terminal galactose may be oxidized at C-6 by galactose-oxidase and the aldehyde group coupled with an activated PEG. In sialoglycoproteins the galactose must be previously uncovered by desialylation. This strategy was employed for the PEGylation of the human thyroid-stimulating hormone (rhTSH, Thyrogen) (Scheme **3**) [25].

Reactive aldehyde groups may also be created by oxidation with periodate. PEGylation with hydrazide-PEG was applied for the terminal residue of a *N*-linked oligosaccharide in ricin A-chain [26] and also for PEGylation of glucose oxidase [27]. In this case, the hydrazone was stabilized by reduction with cyanoborohydride to afford a more stable chemical bond. The exocyclic chains of sialic acid can be PEGylated with this methodology using milder periodate

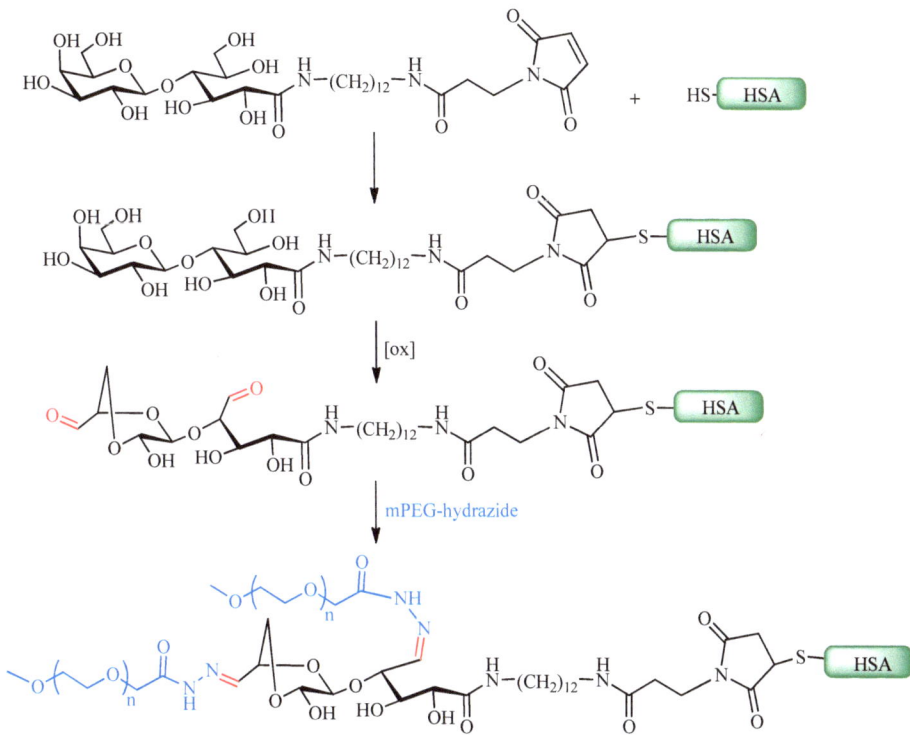

Scheme 2. Site-specific PEGylation of human serum albumin (HSA) at the thiol group of Cys[34]. Adapted from [24].

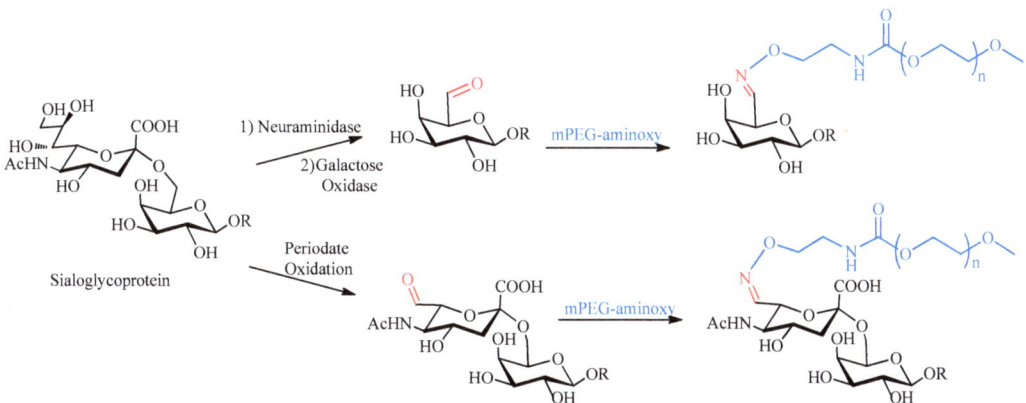

Scheme 3. Strategies for PEGylation of thyrogen at terminal galactose residues or terminal sialic acids. Adapted from [25].

oxidation conditions to create the reactive carbonyl group. Terminal sialic acids from thyrogen were oxidized and later PEGylated with an aminoxy-PEG (Scheme 3) [25].

Localization of O-GlcNAc Glycosylation in a Protein by PEGylation

Clark *et al.* [37, 48] described a method based on PEGylation of the sugar, for the visualization and quantification of *O*-linked *N*-acetylglucosamine units in glycoproteins. The method involves enzymatic glycosylation of the *O*-GlcNAc residues *in vitro* with an unnatural ketogalactose by a mutant galacto-syltransferase, followed by conjugation of the ketose group with an aminoxy-functionalized PEG (Scheme **4a**). PEG functions as a mass tag that shifts the molecular weight of the *O*-glycosylated proteins by an amount that depends on the PEG used. The modified subpopulations are resolved by SDS-PAGE and identified by immunoblotting. A variation on this approach replaces the ketogalactose by an *N*-2-azidoacetylgalactosamine suitable for a click-reaction with an alkyne-functionalized PEG (Scheme **4b**).

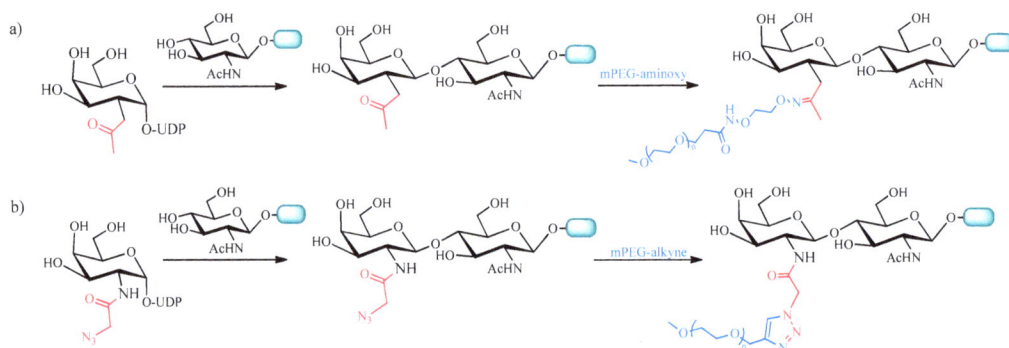

Scheme 4. Enzymatic glycosylation of *O*-GlcNAc in glycoproteins with (**a**) a ketogalactose or with (**b**) *N*-2-azidoacetylgalactosamine. Adapted from [37].

More recently, Teo *et al.* [49] proposed a different strategy based on the Bertozzi's method for *in vivo* incorporation of an *N*-2-azidoacetylglucosamine (GlcNAz) in the *O*-GlcNAc glycosylation sites of a protein. They used peracetylated *N*-2-azidoacetylgalactosamine (Ac4GalNAz) which is catabolized *in vivo* to UDP-GlcNAz, a donor substrate for GlcNAz [50]. The azido-sugar was tagged by a "click" reaction with azadibenzylcyclooctyne (DBCO, also known as ADIBO) available as a PEG5kD conjugate (DBCO-PEG5k) from commercial sources (Scheme **5**). The PEGylated glycoproteins could be analyzed by SDS-PAGE.

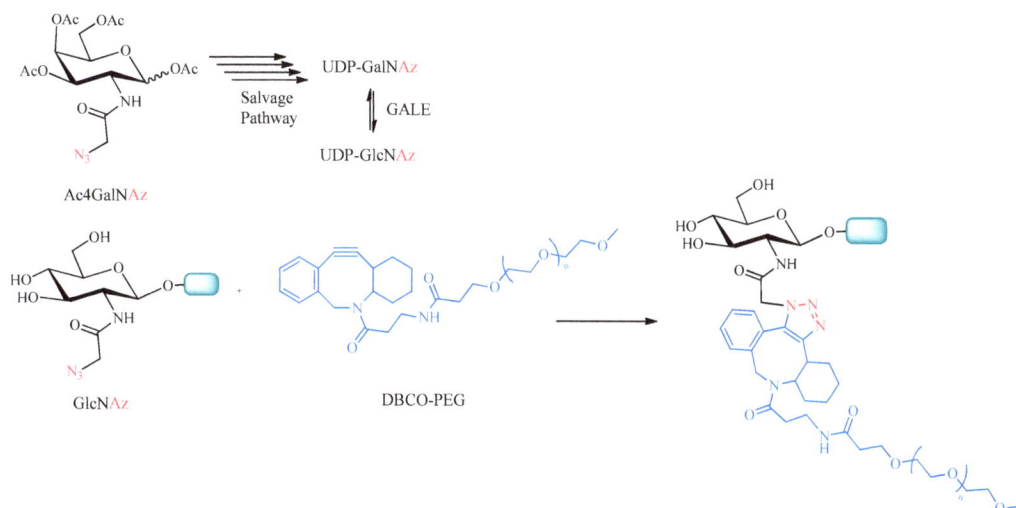

Scheme 5. *In vivo* incorporation of GlcNAz into glycoproteins and further reaction with DBCO-PEG.

PEGYLATION FOR GLYCAN DELIVERY

Carbohydrates are candidates for chemotherapy because they participate in molecular recognition events such as host-pathogen interactions and are responsible for mammal infections [51]. However, there are few reports on the application of PEGylation in carbohydrate-based drugs.

Small Sugar-Based Drugs

The expression level of glucose in different tissues depends on its metabolic consumption, being higher in cancerous cells than in normal ones. This phenomenon, together with an increase in the glucose transporter GluT1 activity was observed in several cancer cell types [52, 53]. Taking advantage of the dependence of cancer cells on glucose to meet their metabolic requirements, Narayanan *et al.* [54] synthesized a PEGylated glucose compound (PEG-Glc) which, by binding to GluT1, blocks glucose uptake by these cells. The synthesis of PEG-Glc is presented in Scheme **6**. Glucose was coupled with a 4-arm PEG

(BrP) derivatized with amino terminal groups, *via* imine formation followed by hydrogenation. Non-functionalized amino groups were capped by amidation with caprolactone (CL). The resulting product was designated as glucose-conjugated branched PEG (GBrP). Although the PEG-NH$_2$ used has three more reactive groups the authors obtained monosubstitution by using equimolar amounts of reactants. The GBrP significantly inhibited the viability of cancer cells and diminished cancer development as compared with a control group.

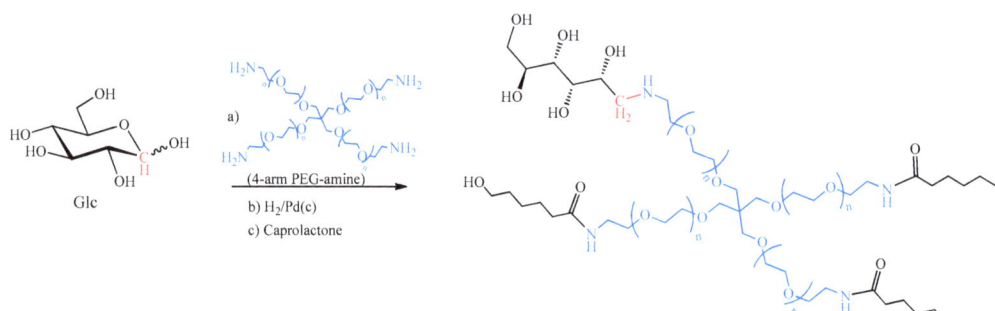

Scheme 6. Conjugation of glucose with a branched PEG (GBrP). Adapted from [54].

Lactose and analogs have been reported as inhibitors of the *Trypanosoma cruzi* trans-sialidase (TcTS), an enzyme that plays a key role in the biology of the parasite [55 - 57]. In fact lactitol inhibited TcTS reaction *in vitro* by being a preferential sialyl acceptor during the transfer reaction, and effectively interfered with parasite infection in cultured cells [58]. Due to its short half-life in blood [59], covalent conjugation of lactitol and analogues with PEG has been developed. Three approaches were described [60] (Scheme **7**): In the first one, lactose and a PEG amine were coupled by reductive amination with NaBH$_3$CN. In the second case, taking advantage of lactobionolactone reactivity towards nucleophiles, amidation was performed by direct incubation with a PEG amine. In the third approach, the amino group was provided by benzyl-β-D-galactopyranosyl-(1→6)-2-amino-2-deoxy-α-D-glucopyranoside and the carboxylic acid by a NHS-activated PEG. These compounds have proved to inhibit TcTS but no significant increase in the permanence in blood was observed. However, improved bioavailability with retention of inhibition of TcTS was achieved by PEGylation with an eight-arm PEG of MW40000. In this case, using an excess of sugar, a product carrying sugar units at the 8 arms could be isolated and characterized (Scheme **8**) [61].

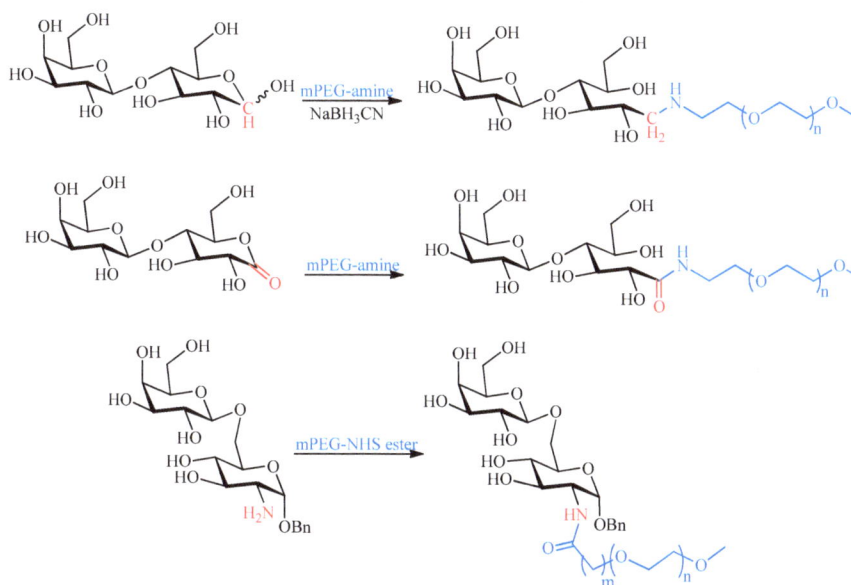

Scheme 7. PEGylation of lactose analogs for TcTS inhibition. Adapted from [60].

Polysaccharidic Drugs

Radix Ophiopogonis polysaccharide (ROP), a natural graminan-type fructan with MW of ~5 kDa, has been found to have an excellent anti-myocardial ischemic activity. However, its rapid renal excretion following administration remarkably limits its efficacy and clinical use, which makes necessary the development of an effective delivery system. In order to improve its pharmacological action and reduce the administration frequency, mPEG-ROP conjugates were synthesized through a coupling reaction between the hydroxyl-activated ROP and an amino-terminated mPEG (Scheme **9**). Probably, substitution at the primary hydroxyl groups of the fructan also occurred, yielding a heterogeneous product. Hydroxyl-activated ROP was prepared by reaction with p-nitrophenyl chloroformate [62].

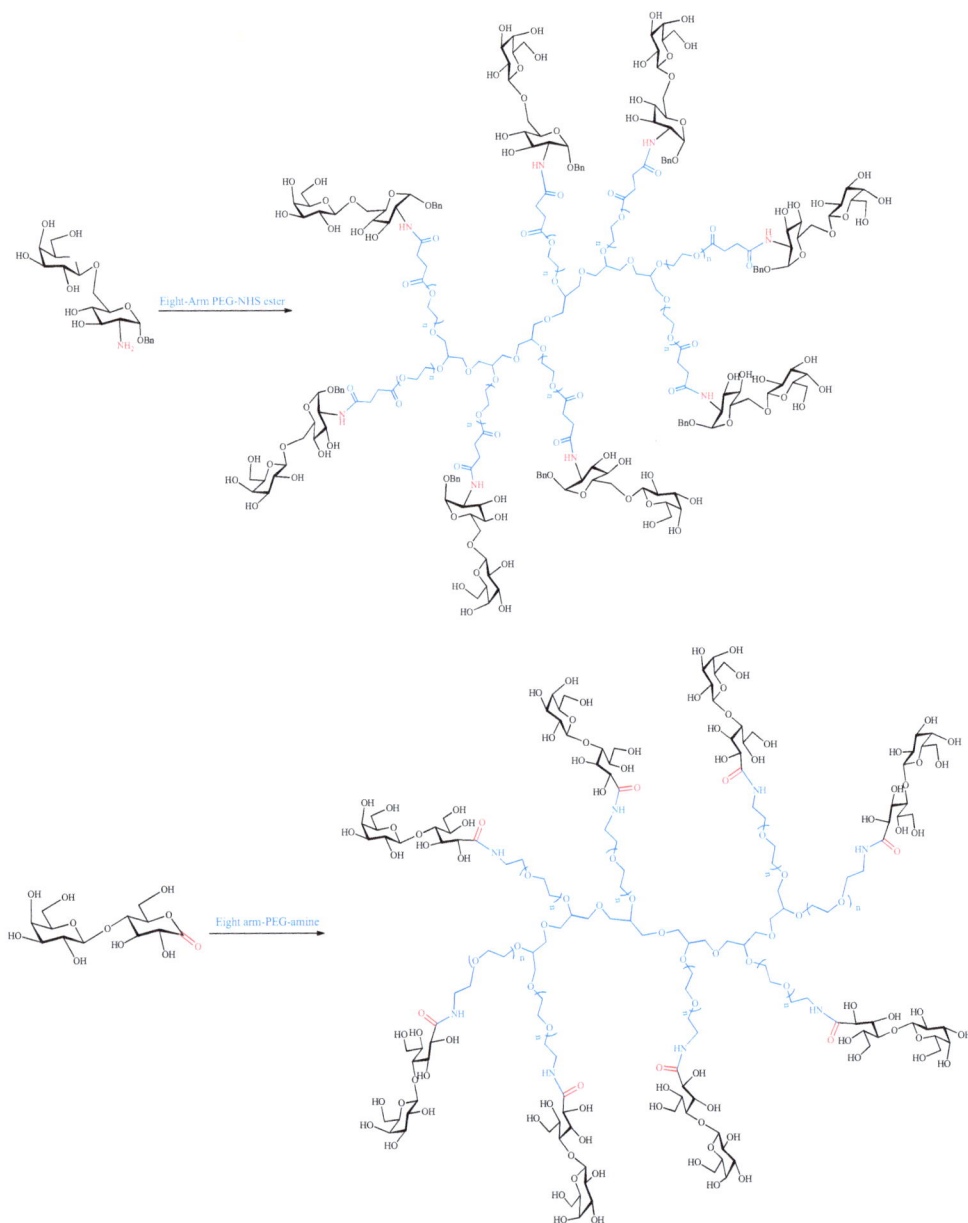

Scheme 8. PEGylation of lactose analogs with multiarm-PEGs. Adapted from [61].

The conjugate, with an average of 1.3 mPEG residues (20 kDa) per single ROP was found to be satisfactory both in plasma retention and in bioactivity assays. It had a 47.4-fold increased half-life and preserved approximately 74% of the bioactivity of ROP. When comparing mono-PEGylated-ROP conjugates with ROP loading in long-circulating liposomes (L-Lps) both were effective in passively targeting the drug to ischemic myocardium, although the latter appeared to induce stronger effects [63].

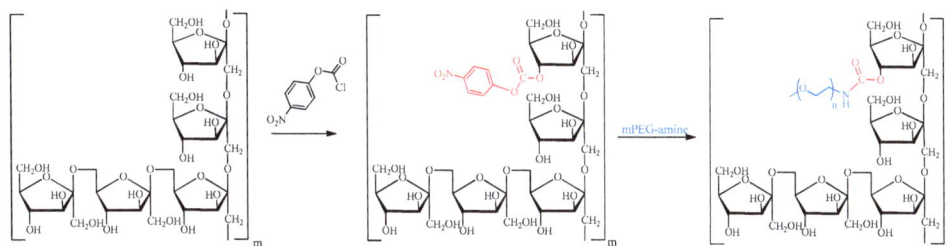

Scheme 9. PEGylation of Radix Ophiopogonis polysaccharide. Adapted from [62].

PEGYLATED POLYSACCHARIDES FOR DRUG/GENE DELIVERY

Most nanocarrier drug/gene delivery systems consist of a macromolecule conjugated through a linker to a drug or gene. Among the various delivery agents tested, sugar-containing polymers have demonstrated great potential. In this respect, chitosan, schizophyllan, curdlan, dextran and carboxymethyldextran, among others, have been conjugated with PEG polymers in order to increase their solubility, reduce immunogenicity and antigenicity and prevent or minimize enzymatic degradation. Conjugation methods include non-covalent interactions and *graft* or *block* covalent copolymerization.

Chitosan

Chitosan (CS) is easily obtained from chitin by alkaline or enzymatic degradation. The sugar backbone consists of β1→4 linked D-glucosamine with a variable degree of N-acetylation. Acetylated and non-acetylated glucosamine residues can be arranged in a random distribution. The most important problem related to the use of chitosan for drug delivery is its limited solubility at neutral and alkaline pH values. Under acidic conditions protonation of the amino groups of the D-glucosamine units allows its solubilization. PEGylation increased chitosan solubility in a broad pH range.

Covalent grafting has been carried out at the NH$_2$ groups by different approaches summarized in Scheme **10** [64]. They include: a) reaction with PEG-aldehyde followed by NaBH$_3$CN reduction or hydrogenation [65 - 67]; b) cross-linking using a diepoxide terminated PEG [68]; c) amide or carbamate formation by reaction with PEG-NHS or PEG-p-nitrophenyl carbonate [69 - 76]; d) amidation by reaction with PEG-COOH [77 - 79]; e) *N*-alkylation by SN$_2$ displacement on a PEG-I [80]; f) "click" reaction between an azido-chitosan and an alkyne-derivatized PEG [81]; g) direct substitution at the 6-OH with a PEG-I after protection of the NH$_2$-groups in chitosan [82, 83]; h) oxidation of C-6 to aldehyde followed by reductive amination with PEG-NH$_2$/ NaBH$_3$CN [84]; i) regioselective esterification with a PEG-COOH [85]. A one-pot reaction has also been performed for the simultaneous *O*- and *N*-PEGylation of chitosan [86].

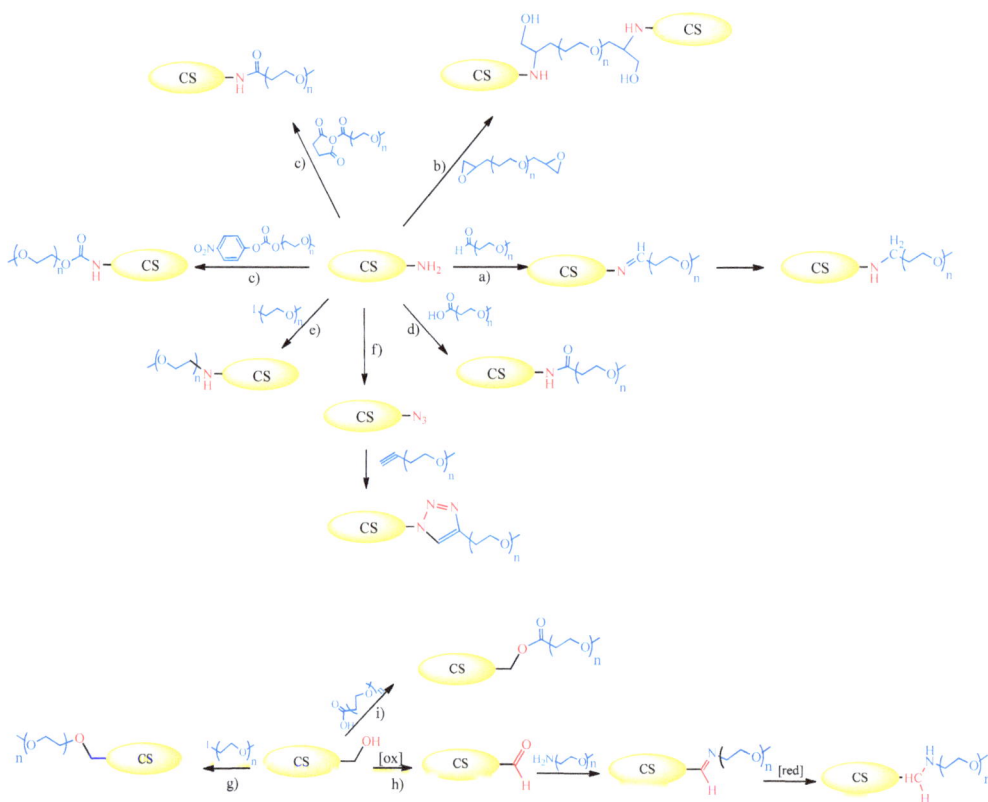

Scheme 10. Strategies for PEGylation of chitosan (CS).

PEGylated chitosan has been extensively used for the delivery of drugs or gene material. In this respect a (PEG)-grafted chitosan/DNA complex was efficiently utilized as a gene carrier to liver cells [76]. Another graft copolymer was obtained by direct amidation of chitosan with a PEG-COOH with the purpose of encapsulating all-trans retinoic acid (ATRA). The resulting ATRA-containing micelles inhibited tumor cell migration better than free ATRA [79]. Furthermore, *N*-acylation of the C2-NH$_2$ groups with phthalic anhydride and further PEGylation of the non-sustituted NH$_2$ groups increased the ATRA incorporation efficiency, the stability of the drug-loaded micelles and the drug release behavior [87].

Chitosan PEGylated using PEG-NHS was combined with hyaluronic acid to obtain a versatile nanoparticle delivery system suitable for the transportation of different types of gene materials, *i.e.* pDNA and small interfering RNA (siRNA) [73].

Also, chitosan was chemically modified by grafting a PEG polymer into its C6-OH and lactobionic acid, a receptor ligand, into C2-NH$_2$ position. The resulting nanoparticle (NP) was used for the targeted delivery of pDNA to hepatic cancer cells [83].

By prolonging the circulation half-life, PEGylation also prolongs drug exposure to healthy tissues thereby causing possible side effects. Non-covalent PEGylation offers a more easily sheddable coating, providing the balance between PEGylation and de-PEGylation needed to produce a potent and safe formulation. Corbet *et al.* [88] developed a non-covalent PEGylated chitosan-based nanoparticle loaded with siRNA, for targeting the lactate transporter MCT1 and the glutamine transporter ASCT2, key transporters of energy fuels for cancer cells. The non-covalent association is due to hydrogen bonding and electrostatic cohesive forces between both polymers and was obtained by mixing them in acetate buffer.

Dextran

Dextran, a natural polymer of glucose linked mainly by α1→6-bonds, has also been used as a drug carrier for a long time. However, although dextran-bound drugs have increased circulation times, they could be degraded by dextranases found in many body tissues. PEGylation of dextran to obtain graft-copolymers was performed using one of the following strategies:

1. Activation of some of the hydroxyl groups of dextran with carbonyldiimidazole followed by coupling with ethylenediamine. A heterogeneous product, resulting from the reaction of the different hydroxyl groups, would be obtained. The aminodextran was further PEGylated with PEG succinimidyl propionate (PEG-SPA) (Scheme **11**) [89]. The dextran core, grafted with PEG residues, still bear free reactive groups for attachment of drugs, diagnostic agents, ligands, *etc*.

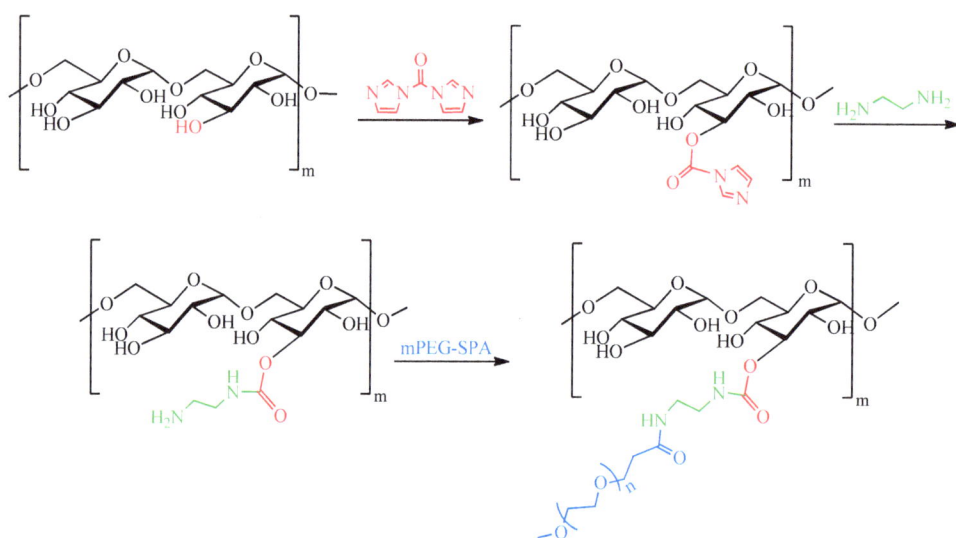

Scheme 11. PEGylation of Aminodextran. Adapted from [89].

2. In the strategy designed by Hosseinkhani *et al.* [90], dextran was partially oxidized with potassium periodate and the dialdehyde derivative was allowed to react with tetramine spermine by reductive amination (Scheme **12**). Dextran-spermine cationic polysaccharide was PEGylated by reaction of the terminal amino groups with and activated *p*-nitrophenylcarbonate-PEG (PEG-NPC). The resulting PEG-dextran-spermine cationic polysaccharide was evaluated as a carrier for the delivery of a gene encoding for β-galactosidase. Spermine was used because polycationic vectors neutralize the negative charge of pDNA, decreasing the electrostatic repulsion with cells and protecting pDNA from digestion by nucleases. More recently, these nanoparticles were use to deliver

pDNA to leukemic cells [91].

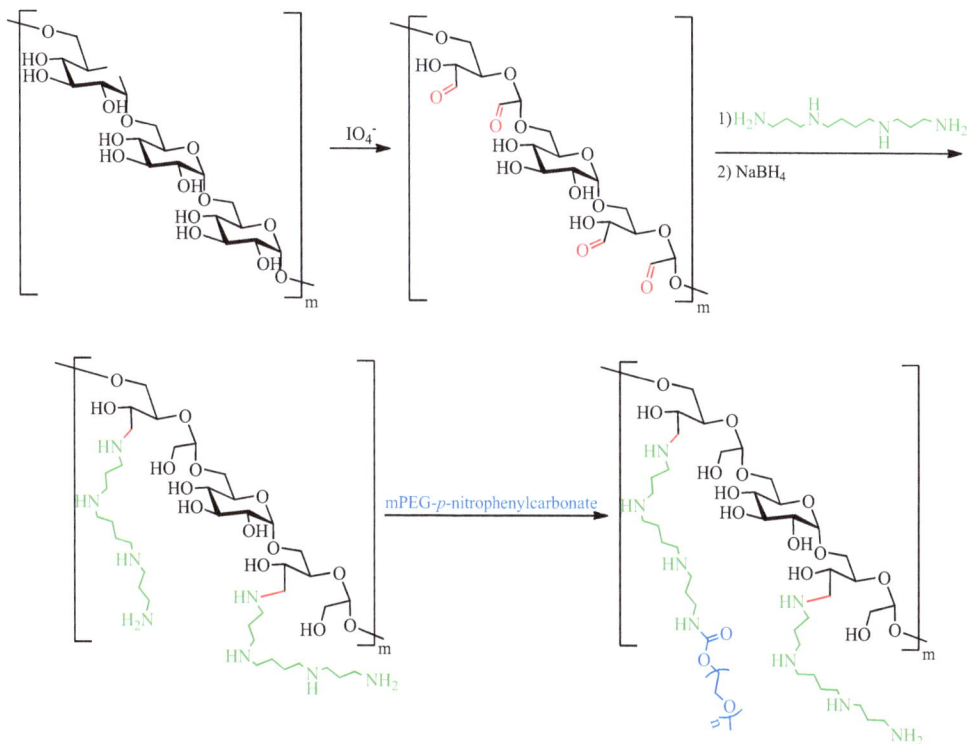

Scheme 12. Synthesis of PEGylated Dextan-spermine cationic polysaccharide. Adapted from [90].

Carboxymethyldextran (CMD)

Carboxymethyldextran (CMD)-*block*-PEG copolymer was prepared by end-to-end coupling of the two biocompatible water-soluble homopolymers, dextran and ω-amino poly(ethylene glycol) (PEG-NH$_2$) [92]. The synthesis involved, in the first place, specific oxidation of the dextran reducing end followed by covalent linkage with a PEG-NH$_2$ *via* a lactone aminolysis reaction. The diblock copolymer dextran-PEG (DEX-PEG) was converted into carboxymethyldextran-PEG (CMD-PEG) by treatment with chloroacetic acid. In Scheme **13** monosubstitution is represented, however, the product would be heterogeneous due to reaction of different hydroxyls and polysubstitution. The CMD-PEG copolymers were specifically designed as substrates with tunable charge density, able to form polyion complex (PIC) micelles upon interaction with a positively charged drug. Therefore, they turned out to be good vehicles for the delivery of cationic drugs like diminazene diaceturate (DIM) [93] and minocycline hydrochloride (MH) [94]. For the delivery of two aminoglycoside antibiotics, paromomycin and neomycin, the cores of the micelles formed by CMD-PEG were strengthened by amidation of the carboxymethyl groups with n-dodecylamine. The introduction of these graft hydrophobic chains increased the non-covalent interactions between hydrophobic groups linked to the charged block of the CMD-PEG [95].

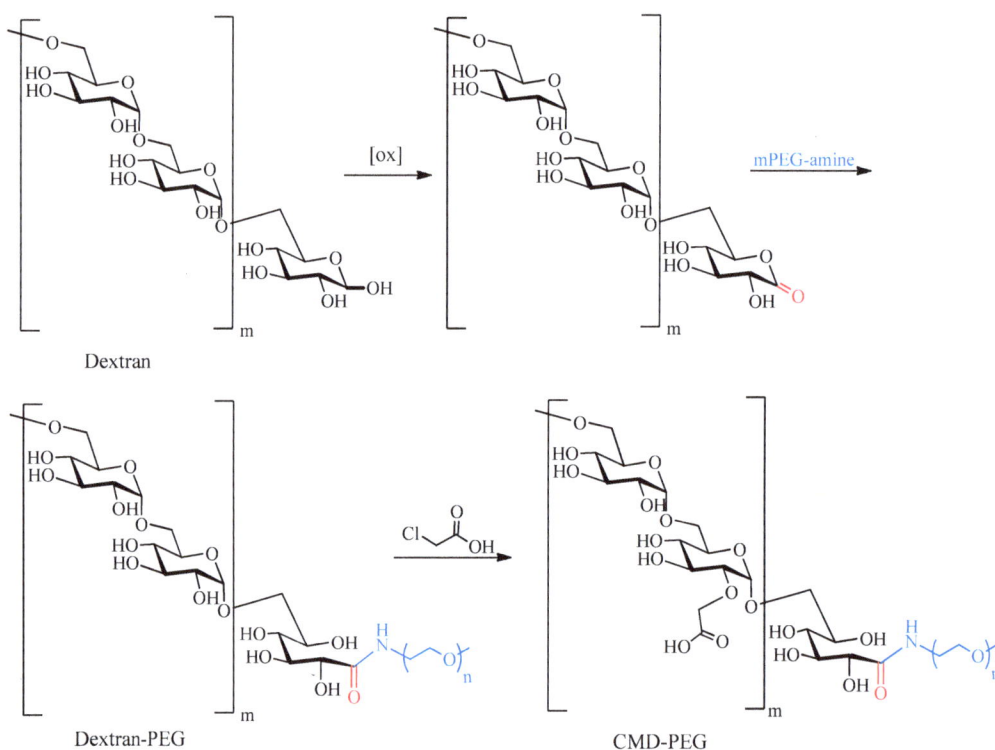

Scheme 13. Synthesis of Dextran-PEG and CMD-PEG.

Glucans

Two (β1→3)-glucans have been used for the synthesis of PEG graft-copolymers with the aim to use them as drug or gene delivery systems. Schizophyllan glucan (SPG) is produced by the fungus *Schizophyllum commune* and consists of a main chain of (β1→3)-linked D-glucoses, branched with one (β1→6)-D-glucosyl unit every three glucose residues. Taking advantage of the resistance of $(1\rightarrow3)$-linked polysaccharides to periodate oxidation, the SPG-graft-PEG copolymer was synthesized by selective oxidation of the terminal glucoses and subsequent reaction of the generated aldehydes with PEG-amine, followed by $NaBH_3CN$ reduction (Scheme **14**). This system was used to efficiently deliver antisense oligonucleotides (AS) designed to suppress the myeloblastosis oncogene c-myb. The SPG-AS-c-myb complex was able to escape from lysosomal degradation [96].

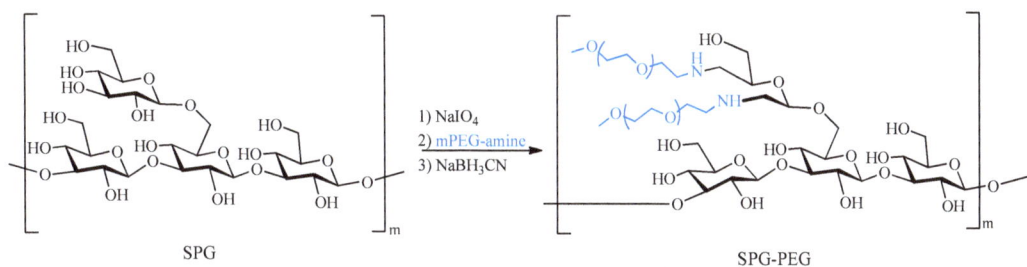

Scheme 14. Synthesis of SPG-graft-PEG copolymer. Adapted from [96].

This technique was slightly modified to introduce a lactose residue in the PEG end for target delivery to the asialoglycoprotein receptor of hepatocytes. Briefly, PEG-diamine (NH_2-PEG-NH_2) was used instead of PEG-NH_2 and the NH_2-PEG-SPG graft copolymer was conjugated with lactobionolactone [97].

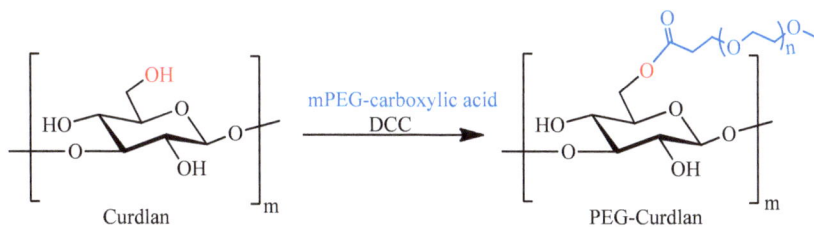

Scheme 15. PEGylation of Curdlan.

Curdlan, on the other hand, is a linear $(\beta1\rightarrow3)$-glucan produced by non-pathogenic bacteria such as *Agrobacterium* sp. The grafting was carried out by dimethylaminopyridine (DMAP)-mediated dicyclohexylcarbodiimide (DCC) ester formation between the C-6 hydroxyls on the curdlan backbone and a terminal carboxylic acid group in PEG-COOH. A core-shell nanoparticle was synthesized by encapsulating the chemotherapeutic drug doxorubicin (DOX) in the curdlan-graft-PEG. The nanoparticle architecture is supposed to consist of the drug DOX covered with the curdlan backbone by hydrophobic interactions, and PEG grafts seem to align outward to provide a water stable shell (Scheme **15**) [98].

In order to use curdlan-graft-PEG in nucleic acid-based drug delivery, particularly siRNA delivery, chemical modifications of natural curdlan, aiming to increase both hydrophilicity and nucleic acid binding affinity, were performed. Replacement of the C-6 hydroxyl groups with amino groups by 6-azido introduction and subsequent reduction gave a cationic 6-amino-6-deoxy-curdlan (6AC). Then, carboxylated mPEG (mPEG-COOH) was grafted to the backbone of 6AC by amide formation with some of the newly introduced C6-amino groups (Scheme **16**) [99]. PEGylated 6AC easily complex with siRNA to form nanoparticles, which can enter human liver cancer cells in culture and substantially knock down endogenous genes such as glyceraldehyde-3-phosphate dehydrogenase (GAPDH) and liver receptor-alpha LXRα.

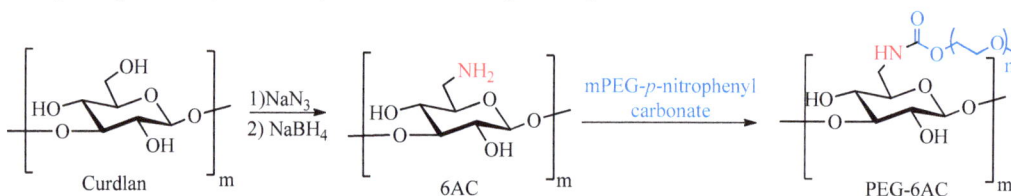

Scheme 16. PEGylation of 6-amino-curdlan. Adapted from [99].

TARGETED DRUG/GENE DELIVERY

The main problems in the development of a new drug are systemic side effects and low bioavailability. In the case of small organic molecule-based drugs, they often have low bioavailability due to limited water solubility and/or poor membrane permeability. Targeted delivery of a drug or gene to its site of action has many therapeutic advantages like increasing therapeutic efficiency while minimizing its systemic toxicity. There are two general types of targeting, passive or active. In passive targeting, the physicochemical properties of the carrier system such as surface charge, hydrophobicity, size and shape should be altered to achieve targeting. Usual modifications include cationic microparticles, surface

PEGylation and use of nanoparticles.

Active targeting is generally performed by loading a macromolecular carrier with an appropriate drug or gene, and by targeting the drug/gene carrier to specific cell or tissue with the help of targeting ligands. Encapsulation protects drugs against the external medium and targeting allows a large percentage of the active compound to reach the pathological area, diminishing the doses and frequency of the intake. In the last years, carbohydrates have been extensively used as targeting moieties since certain cell membrane receptors recognize specific carbohydrate motifs and internalize them *via* endocytosis.

Galactose/Lactose Decorated Target Delivery Systems

The asialoglycoprotein receptor (ASGP-R) is a particularly attractive target due to its very high density on hepatocyte surfaces (50,000-500,000 per cell) and galactose-specific binding property. The receptor recognizes ligands with terminal galactose or *N*-acetylgalactosamine residues and endocytoses the drug system for intracellular degradation. Synthetic ligands with galactosylated or lactosylated residues are significantly efficient at targeting the liver. Several strategies have been used for the β-galactopyranosyl incorporation involving galactose or lactose precursors and diverse glycosylation methods. In this section a brief description of these strategies is summarized.

Use of Galactose or Lactose p-Aminophenyl-β-D-Glycoside

Lactose, as its *p*-aminophenyl glycoside was PEGylated by reductive amination with an aldehyde-ended PEG-poly(2-(dimethylamino)ethyl methacrylate) block copolymer (DMAEMA-PEG-CHO, Scheme **17**) to construct a polyion complex (PIC) micelle-type gene vector for the delivery of plasmid DNA encoding luciferase to hepatoma cells. The lactosylated PIC micelle revealed enhanced transfection compared to the control PIC micelle at a lower dose of pDNA [100].

A similar approach using a bifunctional PEG with a mercapto-terminal group (HS-PEG-Acetal) was used to modify the surface of gold nanoparticles with galactosylated PEG strands with the aim to use them as a colloidal biosensor for galactose-binding lectins (Scheme **18**) [101]. Gold colloids are convenient carriers because they have low cytotoxicity, can be synthesized with defined sizes and can be surface-modified by reaction with thiol-containing molecules. Bergen *et al.* [102] used a *p*-aminophenyl glycoside for the conjugation of galactose to a bifunctional PEG derivative containing a NHS ester group and an *O*-pyridyl disulfide group in the opposite end (OPSS-PEG-NHS) (Scheme **19**). They obtained various nanoparticle formulations of different particle size, surface charge, surface hydrophilicity and galactose ligand density showing hepatocyte targeted delivery *in vivo*.

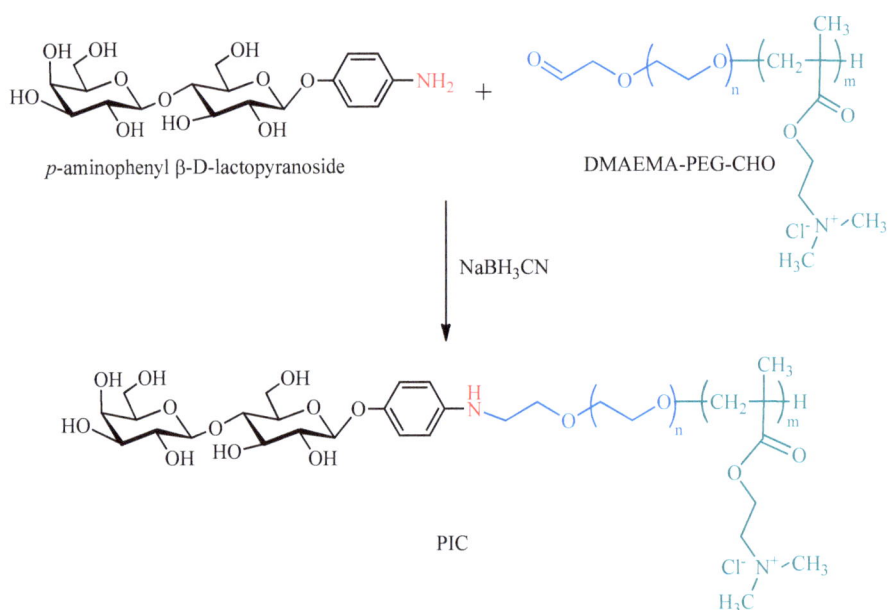

Scheme 17. Synthesis of lactose-conjugated polyion complex micelles.

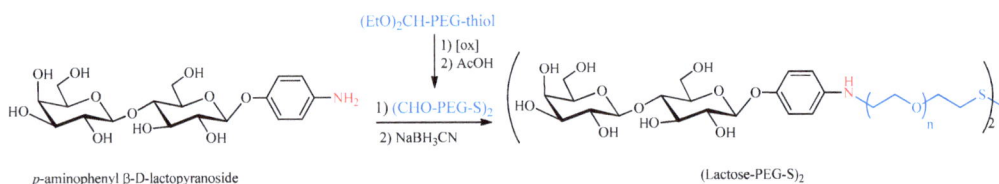

Scheme 18. Synthesis of galactosylated PEG strands for targeted gold nanoparticles. Adapted from [101].

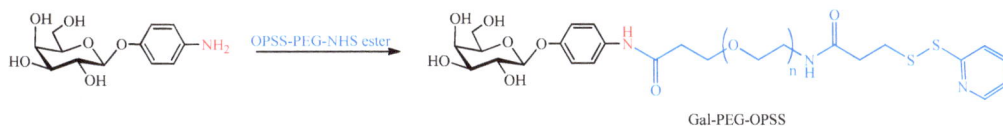

Scheme 19. Synthesis of GAL-PEG-OPSS for incorporation in nanoparticles. Adapted from [102].

In another gene delivery system (polyplexes) to hepatocytes, *p*-aminophenyl β-D-galactopyranoside was conjugated with the NHS group of a bifunctional PEG having an α-vinyl sulfone in the opposite end (VS-PEG-NHS) (Scheme **20**). The vinyl sulfone end was bound with the amino groups of polyethylenimine (PEI) that provide the carrier with the positive charge necessary to interact with plasmid DNA. This polymeric PEI derivative with terminally galactose-grafted PEG, retained the ability to interact with pDNA electrostatically and showed efficient transfection *via* asialoglycoprotein receptors on the surface of hepatoma cells [103].

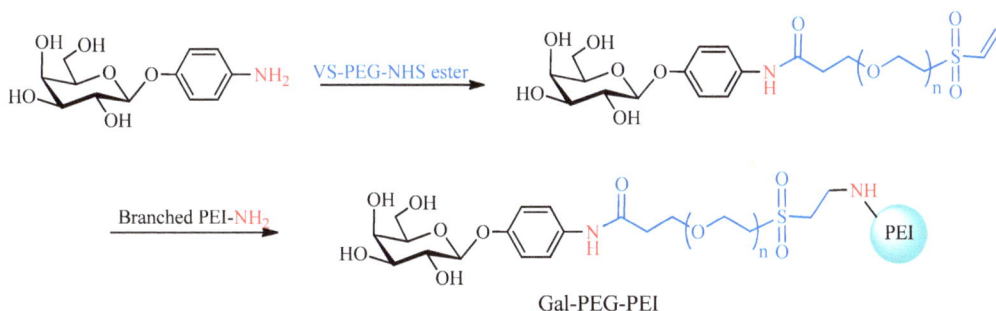

Scheme 20. Synthesis of a galactose-poly(ethylene glycol)-polyethylenimine (Gal-PEG-PEI) for gene delivery to hepatocytes.

Chen *et al.* [104] designed similar polyplexes using another bifunctional PEG derivative with NHS for coupling with the amino group of the *p*-aminophenyl-β-D-galactopyranoside and a maleimide group for PEI-bonding (NHS-PEG-MAL). This system was developed for targeted delivery of pDNA to the β-galactoside-binding lectins abundantly expressed in the lung. The resulting galactose-PEG-PEI/pDNA polyplexes showed improved solubility, stability and reduced toxicity compared with the controls (Scheme **21**).

A pH-sensitive polyion complex (PIC) micelle of siRNA was developed based on the conjugation of siRNA with lactosylated PEG through acid-labile linkage of β-thiopropionate (Lac-PEG-siRNA; Scheme **22**), followed by the complexation with poly(L-lysine). Lac-PEG-siRNA was synthesized by conjugation of 4-aminophenyl lactopyranoside with the carboxyl group of a bifunctional COOH-PEG-acrylate. Michael addition of the Lac-PEG-acrylate toward a 5′-thiol modified sense RNA (ssRNA) followed by annealing with antisense-RNA rendered the Lac-PEG-siRNA. This "smart" polyplex showed a significant and prolonged inhibition of hepatic multicellular tumor spheroids [105].

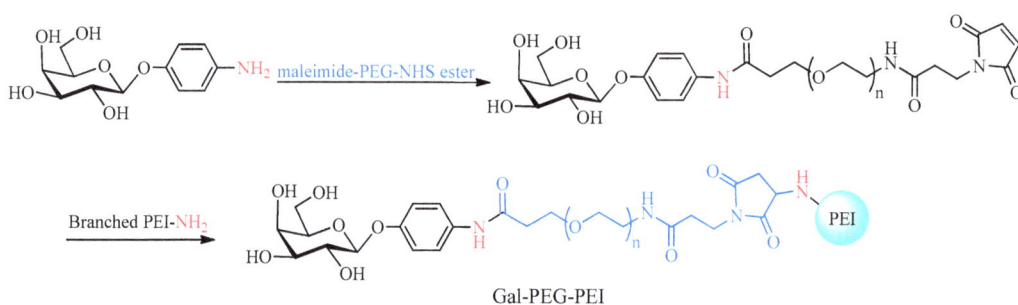

Scheme 21. Synthesis of a galactose-poly(ethylene glycol)-polyethylenimine (Gal-PEG-PEI) for targeted delivery to lungs.

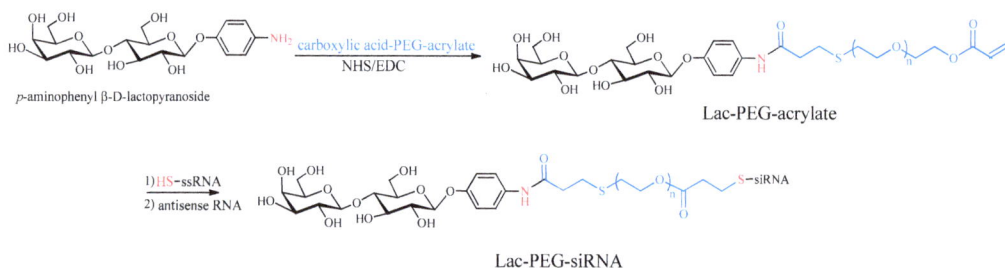

Scheme 22. Synthesis of lactosylated poly(ethylene glycol)-siRNA (Lac-PEG-siRNA) polyplexes.

Use of Lactobionic Acid

The carboxylic group of lactobionic acid is frequently used to incorporate a β-galactopyranosyl residue to the end of a PEG chain through an amide linkage. Biodegradable nanoscopic hydrogels were prepared for their application in targeted drug delivery. The approach was based in the synthesis of a diblock copolymer of PEG and poly(ethyl ethylene phosphate) (PEEP) functionalized in its PEG end with an amino group (NH$_2$-PEG-PEEP-Acr). Reaction with lactobionic acid activated with NHS/EDC yielded the lactosylated nanogel (Lac-PEEP-PEG-Acr, Scheme **23**).

The block copolymer is soluble in water but self-assembles into core-shell structural nanoparticles upon addition of salt. After UV-crosslinking to fix the structure and dialysis to remove salts, the nanoparticles become totally hydrophilic, generating nanogel particles with an inner reservoir for water-soluble drugs [106].

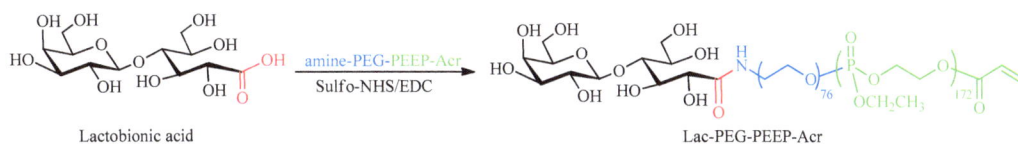

Scheme 23. Synthesis of lactosylated hydrogels for targeted drug delivery.

Gold nanoparticles had also been targeted with β-galactopyranosyl units by functionalization with HS-PEG-NH$_2$ followed by conjugation with lactobionic acid activated with sulfo-NHS/EDC. The obtained nanoparticles enhanced the cytotoxic effects of radiation therapy, concentrating the effect on targeted tumor cells [107].

Use of 2-Amino-2-Deoxy-D-Galactose

The amino group of 2-amino-2-deoxy-D-galactose was used for amide formation with a carboxy-PEG-SH that was also linked to silver nanoparticles. With the aim to use them for the detection of cholera toxin by surface enhanced Raman spectroscopy, the nanoparticles were also modified with sialic acid residues linked through its carboxylic acid to an amino-PEG-SH (Scheme **24**). The mixed-carbohydrate particles mimic the GM1 ganglioside expressed in the surface of the host's intestinal cells that interacts with lectins from the pili of Vibrio cholerae bacteria [108].

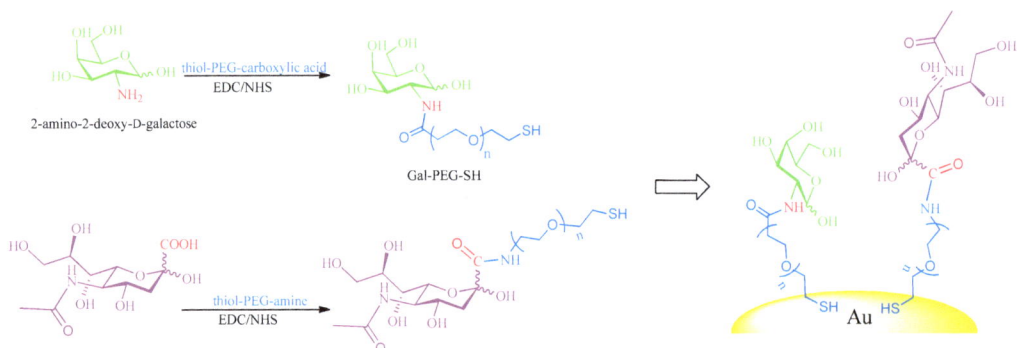

Scheme 24. PEGylation of GalNH$_2$ and sialic acid and schematic representation of nanoparticles mimicking GM1 ganglioside. Adapted from [108].

β-Galactosylation by Other Methods

Galactose was introduced in the distal end of a PEGylated cholesterol by direct glycosylation using the trichoroacetimidate method (Scheme **25**).

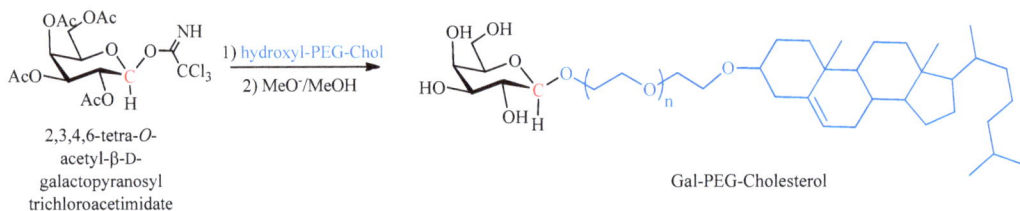

Scheme 25. Synthesis of galactosylated PEG-cholesterol liposomes. Adapted from [109].

The galactosylated cholesterol derivative contains a lipophilic anchor moiety (cholesterol) for stable incorporation into liposomes, a polyethylene glycol for steric stabilization and long circulating effect, and a galactose moiety for targeting the cell surface receptors in hepatocytes. When used for doxorubicin delivery, cytotoxicity assays showed that the cell proliferation inhibition effect of galactose modified liposomes was higher than that of the unmodified liposomes [109].

β-D-galactopyranose could also be introduced by reaction of thiogalactose with a protected bromoacetic acid followed by amidation with bifunctional amino-PEG-azido (Scheme **26**). The azide-end underwent a "click reaction" with several branches of an alkyne-functionalized 12-arm dendrimer designed as a doxorubicin vehicle to hepatoma cells [110].

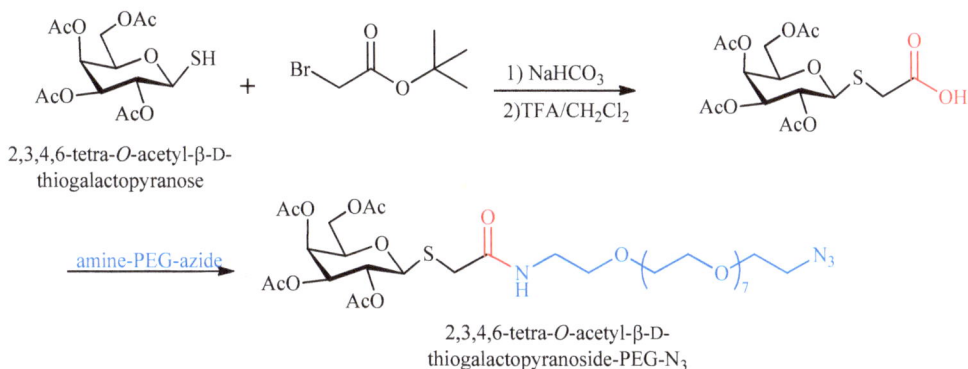

Scheme 26. Synthesis of PEGylated thiogalactopyranoside for "click reaction" with dendrimers.

Mannose-Decorated Target Delivery Systems

Macrophages play an important role in a wide range of diseases, engulfing and digesting cellular debris, foreign substances, microbes, cancer cells, *etc.* They accumulate at pathological sites and have been an important target for drug delivery. The inclusion in a drug of specific ligands for the macrophage receptor(s) significantly enhances the rate and extent of uptake by the macrophages. Mannose receptors are abundantly expressed in liver, spleen, and alveolar macrophages [111]. Strategies for the introduction of a mannose moiety in a drug or gene carrier are similar to those described above for galactose-targeted systems.

p-aminophenyl-α-D-mannopyranoside was used for the introduction of a mannose residue in the surface of a siRNA-containing cationic cyclodextrin nanoparticle (Scheme **27**) [112].

The nanoparticle formulation was based on electrostatic interactions between siRNA and a cationic cyclodextrin-containing polymer targeted to the kidney by Man-PEG-adamantine (AD) chains. Adamantine is known to form inclusion complexes and supramolecular oligomeric assemblies with cyclodextrin. For the synthesis of Man-PEG-AD, *p*-aminophenyl-α-D-mannopyranoside was reacted with the NHS end of a bifunctional PEG, containing a maleimide group in the other end (NHS-PEG-MAL), suitable for Michael addition of a SH-containing adamantine molecule. The mannose ligand enhanced the *in vitro* uptake of the siRNA/CDP-NPs in mouse and human mesangial cells.

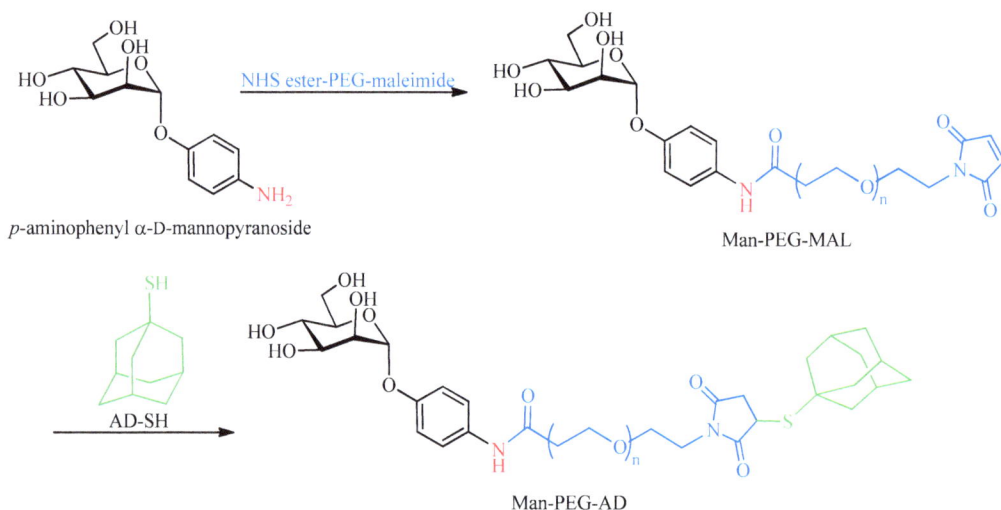

Scheme 27. Synthesis of Man-PEG-AD for the construction of a siRNA delivery nanoparticle.

Other glycosides were also used for the attachment of a mannopyranosyl unit. Among them, 2-imino-2-methoxyethyl-1-thiomannoside (IME-thiomannoside) reacted with the amino end of an NH_2-PEG-distearoylphosphatidylethanolamine (DSPE) to obtain a mannosylated liposome. Further introduction of pDNA and perfluoropropane imaging gas, generated an ultrasound (US)-responsive and mannose modified gene carrier, named Man-PEG bubble lipoplex (Scheme **28**). In this gene transfection method, pDNA can directly introduce the nucleic acids into the cells through the transient pores created by US-responsive degradation of bubble lipoplexes [113].

More recently, these Man-PEG bubble lipoplexes, carrying pDNA were administered to mice previously injected with doxorubicin-encapsulated PEGylated liposomes, in combination with ultrasound irradiation for the treatment of melanoma [114].

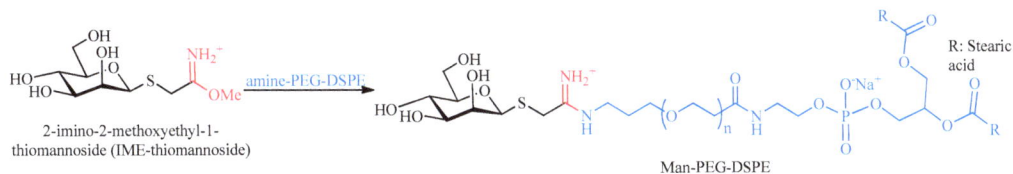

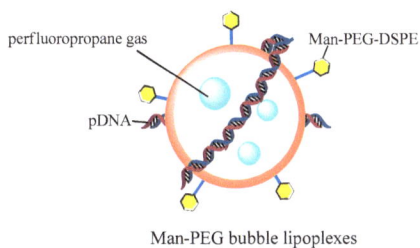

Scheme 28. Synthesis of a mannosylated liposome and schematic representation of the bubble lipoplex. Adapted from [113].

CONCLUDING REMARKS

PEGylation has been mainly applied to improve the pharmacokinetic properties of proteins without affecting their activity. In some cases, the attenuation of immunogenicity was beneficial, especially when PEGylated viruses were used as gene vectors. Site selective PEGylation by glycoPEGylation yields a more homogeneous product, allowing the therapeutic use of PEGylated drugs, like lipegfilgrastim, already in the market. Besides being used to improve the bioavailability of proteins, carbohydrate PEGs linked to a carrier are being used for active targeting to cell receptors for sugars. Most of the drug delivery systems were designed for the asialoglycoprotein receptor in mammal cells and examples of their preparation are illustrated in this review. It is expected that future studies will be extended to active targeting to receptors for other sugars and other organisms like pathogenic microorganisms.

CONFLICT OF INTEREST

The authors confirm that they have no conflict of interest to declare for this publication.

ACKNOWLEDGEMENTS

This work was supported by grants from Agencia Nacional de Promoción Científica y Tecnológica (ANPCyT), Consejo Nacional de Investigaciones Científicas y Técnicas (CONICET) and Universidad de Buenos Aires. R Agustí, ME Giorgi and R M de Lederkremer are research members of CONICET.

REFERENCES

[1]   Smolinske SC. Handbook of food, drug and cosmetic excipients. Boca Raton: CRC Press 1992; pp. 287-94.

[2]   Veronese FM, Pasut G. PEGylation, successful approach to drug delivery. Drug Discov Today 2005; 10(21): 1451-8.
       [http://dx.doi.org/10.1016/S1359-6446(05)03575-0] [PMID: 16243265]

[3]   Abuchowski A, van Es T, Palczuk NC, Davis FF. Alteration of immunological properties of bovine serum albumin by covalent attachment of polyethylene glycol. J Biol Chem 1977; 252(11): 3578-81.
       [PMID: 405385]

[4]   Jevsevar S, Kunstelj M, Porekar VG. PEGylation of therapeutic proteins. Biotechnol J 2010; 5(1): 113-28.
       [http://dx.doi.org/10.1002/biot.200900218] [PMID: 20069580]

[5]   Piedmonte DM, Treuheit MJ. Formulation of Neulasta (pegfilgrastim). Adv Drug Deliv Rev 2008; 60(1): 50-8.
       [http://dx.doi.org/10.1016/j.addr.2007.04.017] [PMID: 17822802]

[6]   Harris JM, Chess RB. Effect of pegylation on pharmaceuticals. Nat Rev Drug Discov 2003; 2(3): 214-21.
       [http://dx.doi.org/10.1038/nrd1033] [PMID: 12612647]

[7]   Bailon P, Won CY. PEG-modified biopharmaceuticals. Expert Opin Drug Deliv 2009; 6(1): 1-16.
       [http://dx.doi.org/10.1517/17425240802650568] [PMID: 19236204]

[8]   Blick SK, Curran MP. Certolizumab pegol: in Crohns disease. BioDrugs 2007; 21(3): 195-201.
       [http://dx.doi.org/10.2165/00063030-200721030-00006] [PMID: 17516714]

[9]   Sato H. Enzymatic procedure for site-specific pegylation of proteins. Adv Drug Deliv Rev 2002; 54(4): 487-504.
       [http://dx.doi.org/10.1016/S0169-409X(02)00024-8] [PMID: 12052711]

[10]   Scaramuzza S, Tonon G, Olianas A, *et al*. A new site-specific monoPEGylated filgrastim derivative prepared by enzymatic conjugation: Production and physicochemical characterization. J Control Release 2012; 164(3): 355-63.
       [http://dx.doi.org/10.1016/j.jconrel.2012.06.026] [PMID: 22735238]

[11]   Mishra P, Nayak B, Dey RK. PEGylation in anti-cancer therapy: An overview. Asian J Pharm Sci 2016; 11(3): 337-48.

[12]   Levy Y, Hershfield MS, Fernandez-Mejia C, *et al*. Adenosine deaminase deficiency with late onset of

recurrent infections: response to treatment with polyethylene glycol-modified adenosine deaminase. J Pediatr 1988; 113(2): 312-7.
[http://dx.doi.org/10.1016/S0022-3476(88)80271-3] [PMID: 3260944]

[13]   Graham ML. Pegaspargase: a review of clinical studies. Adv Drug Deliv Rev 2003; 55(10): 1293-302.
[http://dx.doi.org/10.1016/S0169-409X(03)00110-8] [PMID: 14499708]

[14]   Wang YS, Youngster S, Grace M, Bausch J, Bordens R, Wyss DF. Structural and biological characterization of pegylated recombinant interferon alpha-2b and its therapeutic implications. Adv Drug Deliv Rev 2002; 54(4): 547-70.
[http://dx.doi.org/10.1016/S0169-409X(02)00027-3] [PMID: 12052714]

[15]   Rajender Reddy K, Modi MW, Pedder S. Use of peginterferon alfa-2a (40 KD) (Pegasys) for the treatment of hepatitis C. Adv Drug Deliv Rev 2002; 54(4): 571-86.
[http://dx.doi.org/10.1016/S0169-409X(02)00028-5] [PMID: 12052715]

[16]   Kinstler O, Molineux G, Treuheit M, Ladd D, Gegg C. Mono-N-terminal poly(ethylene glycol)-protein conjugates. Adv Drug Deliv Rev 2002; 54(4): 477-85.
[http://dx.doi.org/10.1016/S0169-409X(02)00023-6] [PMID: 12052710]

[17]   Trainer PJ, Drake WM, Katznelson L, *et al.* Treatment of acromegaly with the growth hormone-receptor antagonist pegvisomant. N Engl J Med 2000; 342(16): 1171-7.
[http://dx.doi.org/10.1056/NEJM200004203421604] [PMID: 10770982]

[18]   Ng EW, Adamis AP. Anti-VEGF aptamer (pegaptanib) therapy for ocular vascular diseases. Ann N Y Acad Sci 2006; 1082: 151-71.
[http://dx.doi.org/10.1196/annals.1348.062] [PMID: 17145936]

[19]   Macdougall IC. CERA (Continuous Erythropoietin Receptor Activator): a new erythropoiesis-stimulating agent for the treatment of anemia. Curr Hematol Rep 2005; 4(6): 436-40.
[PMID: 16232379]

[20]   Sherman MR, Saifer MG, Perez-Ruiz F. PEG-uricase in the management of treatment-resistant gout and hyperuricemia. Adv Drug Deliv Rev 2008; 60(1): 59-68.
[http://dx.doi.org/10.1016/j.addr.2007.06.011] [PMID: 17826865]

[21]   Woodburn KW, Fong KL, Wilson SD, *et al.* Peginesatide clearance, distribution, metabolism, and excretion in monkeys following intravenous administration. Drug Metab Dispos 2013; 41(4): 774-84.
[http://dx.doi.org/10.1124/dmd.112.048033] [PMID: 23318685]

[22]   Mahlert F, Schmidt K, Allgaier H, Liu R, Müller U, Shen WD. Rational development of Lipegfilgrastim, a novel long-acting granulocyte colony-stimulating factor, using glycopegylation technology. Blood 2013; 122(21): 4853.

[23]   DeFrees S, Wang ZG, Xing R, *et al.* GlycoPEGylation of recombinant therapeutic proteins produced in *Escherichia coli.* Glycobiology 2006; 16(9): 833-43.
[http://dx.doi.org/10.1093/glycob/cwl004] [PMID: 16717104]

[24]   Salmaso S, Semenzato A, Bersani S, Mastrotto F, Scomparin A, Caliceti P. Site-selective protein glycation and PEGylation. Eur Polym J 2008; 44(5): 1378-89.
[http://dx.doi.org/10.1016/j.eurpolymj.2008.02.021]

[25]   Park A, Honey DM, Hou L, *et al.* Carbohydrate-mediated polyethylene glycol conjugation of TSH

improves its pharmacological properties. Endocrinology 2013; 154(3): 1373-83.
[http://dx.doi.org/10.1210/en.2012-2010] [PMID: 23389953]

[26]   Youn YS, Na DH, Yoo SD, Song SC, Lee KC. Carbohydrate-specifically polyethylene glycol-modified ricin A-chain with improved therapeutic potential. Int J Biochem Cell Biol 2005; 37(7): 1525-33.
[http://dx.doi.org/10.1016/j.biocel.2005.01.014] [PMID: 15833282]

[27]   Ritter DW, Roberts JR, McShane MJ. Glycosylation site-targeted PEGylation of glucose oxidase retains native enzymatic activity. Enzyme Microb Technol 2013; 52(4-5): 279-85.
[http://dx.doi.org/10.1016/j.enzmictec.2013.01.004] [PMID: 23540931]

[28]   Scott LC, Yao JC, Benson AB III, *et al.* A phase II study of pegylated-camptothecin (pegamotecan) in the treatment of locally advanced and metastatic gastric and gastro-oesophageal junction adenocarcinoma. Cancer Chemother Pharmacol 2009; 63(2): 363-70.
[http://dx.doi.org/10.1007/s00280-008-0746-2] [PMID: 18398613]

[29]   Choi JS, Jo BW. Enhanced paclitaxel bioavailability after oral administration of pegylated paclitaxel prodrug for oral delivery in rats. Int J Pharm 2004; 280(1-2): 221-7.
[http://dx.doi.org/10.1016/j.ijpharm.2004.05.014] [PMID: 15265561]

[30]   Hoofnagle JH, Seeff LB. Peginterferon and ribavirin for chronic hepatitis C. N Engl J Med 2006; 355(23): 2444-51.
[http://dx.doi.org/10.1056/NEJMct061675] [PMID: 17151366]

[31]   Eto Y, Yoshioka Y, Ishida T, *et al.* Optimized PEGylated adenovirus vector reduces the anti-vector humoral immune response against adenovirus and induces a therapeutic effect against metastatic lung cancer. Biol Pharm Bull 2010; 33(9): 1540-4.
[http://dx.doi.org/10.1248/bpb.33.1540] [PMID: 20823571]

[32]   Ashwell G, Harford J. Carbohydrate-specific receptors of the liver. Annu Rev Biochem 1982; 51: 531-54.
[http://dx.doi.org/10.1146/annurev.bi.51.070182.002531] [PMID: 6287920]

[33]   Pathak A, Vyas SP, Gupta KC. Nano-vectors for efficient liver specific gene transfer. Int J Nanomedicine 2008; 3(1): 31-49.
[PMID: 18488414]

[34]   de Lederkremer RM, Agusti R. Glycobiology of *Trypanosoma cruzi*. Adv Carbohydr Chem Biochem 2009; 62: 311-66.
[http://dx.doi.org/10.1016/S0065-2318(09)00007-9] [PMID: 19501708]

[35]   Freichels H, Jérôme R, Jérôme C. Sugar-labeled and PEGylated (bio)degradable polymers intended for targeted drug delivery systems. Carbohydr Polym 2011; 86(3): 1093-106.
[http://dx.doi.org/10.1016/j.carbpol.2011.06.004]

[36]   Hart GW, Housley MP, Slawson C. Cycling of O-linked beta-N-acetylglucosamine on nucleocytoplasmic proteins. Nature 2007; 446(7139): 1017-22.
[http://dx.doi.org/10.1038/nature05815] [PMID: 17460662]

[37]   Clark PM, Rexach JE, Hsieh-Wilson LC. Visualization of O-GlcNAc glycosylation stoichiometry and dynamics using resolvable poly(ethylene glycol) mass tags. Curr Protoc Chem Biol 2013; 5(4): 281-

302.
[http://dx.doi.org/10.1002/9780470559277.ch130153] [PMID: 24391098]

[38]   Rabuka D. Chemoenzymatic methods for site-specific protein modification. Curr Opin Chem Biol 2010; 14(6): 790-6.
[http://dx.doi.org/10.1016/j.cbpa.2010.09.020] [PMID: 21030291]

[39]   DeFrees S, Zopf D, Bayer RJ, Bowe C, Hakes D, Chen X. GlycoPEGylation methods and protein/peptides produced by the methods. US patent 2004/0132640 A1, 2004.

[40]   DeFrees S, Felo M. Nucleotide sugar purification using membranes. US patent WO2007056191 A2, 2007.

[41]   Scheckermann C, Schmidt K, Abdolzade-Bavil A, *et al.* Lipegfilgrastim: A long-acting, once-pe--cycle, glycopegylated recombinant human filgrastim. J Clin Oncol 2013; 31(suppl; abstr e13548)

[42]   Zündorf I, Dingermann T. PEGylationa well-proven strategy for the improvement of recombinant drugs. Pharmazie 2014; 69(5): 323-6.
[PMID: 24855821]

[43]   Østergaard H, Bjelke JR, Hansen L, *et al.* Prolonged half-life and preserved enzymatic properties of factor IX selectively PEGylated on native N-glycans in the activation peptide. Blood 2011; 118(8): 2333-41.
[http://dx.doi.org/10.1182/blood-2011-02-336172] [PMID: 21700771]

[44]   Stennicke HR, Østergaard H, Bayer RJ, *et al.* Generation and biochemical characterization of glycoPEGylated factor VIIa derivatives. Thromb Haemost 2008; 100(5): 920-8.
[PMID: 18989539]

[45]   DeFrees S, Zopf D, Taudlte S, Scott Willett W, Bayer RJ, Kalo M. One pot desialylation and glycopegylation of therapeutic peptides. US Patent 8911967 B2, 2014.

[46]   Stennicke HR, Kjalke M, Karpf DM, *et al.* A novel B-domain O-glycoPEGylated FVIII (N8-GP) demonstrates full efficacy and prolonged effect in hemophilic mice models. Blood 2013; 121(11): 2108-16.
[http://dx.doi.org/10.1182/blood-2012-01-407494] [PMID: 23335368]

[47]   Klausen NK, Bjorn S, Behrens C, Garibay PW. Pegylated factor VII glycoforms. US Patent 8053410 B2, 2011.

[48]   Rexach JE, Rogers CJ, Yu SH, Tao J, Sun YE, Hsieh-Wilson LC. Quantification of O-glycosylation stoichiometry and dynamics using resolvable mass tags. Nat Chem Biol 2010; 6(9): 645-51.
[http://dx.doi.org/10.1038/nchembio.412] [PMID: 20657584]

[49]   Teo CF, Wells L. Monitoring protein O-linked β-N-acetylglucosamine status *via* metabolic labeling and copper-free click chemistry. Anal Biochem 2014; 464: 70-2.
[http://dx.doi.org/10.1016/j.ab.2014.06.010] [PMID: 24995865]

[50]   Boyce M, Carrico IS, Ganguli AS, *et al.* Metabolic cross-talk allows labeling of O-linked beta--acetylglucosamine-modified proteins *via* the N-acetylgalactosamine salvage pathway. Proc Natl Acad Sci USA 2011; 108(8): 3141-6.
[http://dx.doi.org/10.1073/pnas.1010045108] [PMID: 21300897]

[51]   Kawasaki N, Itoh S, Hashii N, *et al.* The significance of glycosylation analysis in development of biopharmaceuticals. Biol Pharm Bull 2009; 32(5): 796-800.
[http://dx.doi.org/10.1248/bpb.32.796] [PMID: 19420744]

[52]   Flier JS, Mueckler MM, Usher P, Lodish HF. Elevated levels of glucose transport and transporter messenger RNA are induced by *ras* or *src* oncogenes. Science 1987; 235(4795): 1492-5.
[http://dx.doi.org/10.1126/science.3103217] [PMID: 3103217]

[53]   Macheda ML, Rogers S, Best JD. Molecular and cellular regulation of glucose transporter (GLUT) proteins in cancer. J Cell Physiol 2005; 202(3): 654-62.
[http://dx.doi.org/10.1002/jcp.20166] [PMID: 15389572]

[54]   Narayanan K, Erathodiyil N, Gopalan B, Chong S, Wan AC, Ying JY. Targeting Warburg effect in cancers with PEGylated glucose. Adv Healthc Mater 2016; 5(6): 696-701.
[http://dx.doi.org/10.1002/adhm.201500613] [PMID: 26792539]

[55]   Frasch AC. Functional diversity in the trans-sialidase and mucin families in *Trypanosoma cruzi*. Parasitol Today (Regul Ed) 2000; 16(7): 282-6.
[http://dx.doi.org/10.1016/S0169-4758(00)01698-7] [PMID: 10858646]

[56]   Tomlinson S, Pontes de Carvalho LC, Vandekerckhove F, Nussenzweig V. Role of sialic acid in the resistance of *Trypanosoma cruzi* trypomastigotes to complement. J Immunol 1994; 153(7): 3141-7.
[PMID: 8089492]

[57]   Pereira-Chioccola VL, Acosta-Serrano A, Correia de Almeida I, *et al.* Mucin-like molecules form a negatively charged coat that protects *Trypanosoma cruzi* trypomastigotes from killing by human anti-alpha-galactosyl antibodies. J Cell Sci 2000; 113(Pt 7): 1299-307.
[PMID: 10704380]

[58]   Agustí R, París G, Ratier L, Frasch AC, de Lederkremer RM. Lactose derivatives are inhibitors of *Trypanosoma cruzi* trans-sialidase activity toward conventional substrates *in vitro* and *in vivo*. Glycobiology 2004; 14(7): 659-70.
[http://dx.doi.org/10.1093/glycob/cwh079] [PMID: 15070857]

[59]   Mucci J, Risso MG, Leguizamón MS, Frasch AC, Campetella O. The *trans*-sialidase from *Trypanosoma cruzi* triggers apoptosis by target cell sialylation. Cell Microbiol 2006; 8(7): 1086-95.
[http://dx.doi.org/10.1111/j.1462-5822.2006.00689.x] [PMID: 16819962]

[60]   Giorgi ME, Ratier L, Agusti R, Frasch AC, de Lederkremer RM. Synthesis of PEGylated lactose analogs for inhibition studies on *T.cruzi* trans-sialidase. Glycoconj J 2010; 27(5): 549-59.
[http://dx.doi.org/10.1007/s10719-010-9300-7] [PMID: 20645127]

[61]   Giorgi ME, Ratier L, Agusti R, Frasch AC, de Lederkremer RM. Improved bioavailability of inhibitors of *Trypanosoma cruzi trans*-sialidase: PEGylation of lactose analogs with multiarm polyethyleneglycol. Glycobiology 2012; 22(10): 1363-73.
[http://dx.doi.org/10.1093/glycob/cws091] [PMID: 22653661]

[62]   Lin X, Wang S, Jiang Y, *et al.* Poly(ethylene glycol)-radix Ophiopogonis polysaccharide conjugates: preparation, characterization, pharmacokinetics and *in vitro* bioactivity. Eur J Pharm Biopharm 2010; 76(2): 230-7.
[http://dx.doi.org/10.1016/j.ejpb.2010.07.003] [PMID: 20633648]

[63] Wang L, Yao C, Wu F, Lin X, Shen L, Feng Y. Targeting delivery of Radix Ophiopogonis polysaccharide to ischemic/reperfused rat myocardium by long-circulating macromolecular and liposomal carriers. Int J Nanomedicine 2015; 10: 5729-37.
[PMID: 26425081]

[64] Casettari L, Vllasaliu D, Castagnino E, Stolnik S, Howdle S, Illum L. PEGylated chitosan derivatives: Synthesis, characterizations and pharmaceutical applications. Prog Polym Sci 2012; 37(5): 659-85.
[http://dx.doi.org/10.1016/j.progpolymsci.2011.10.001]

[65] Harris JM, Struck EC, Case MG, *et al.* Synthesis and characterization of Poly(ethylene glycol) derivatives. J Polym Sci Polym Chem 1984; 22(2): 341-52.

[66] Sugimoto M, Morimoto M, Sashiwa H, Saimoto H, Shigemasa Y. Preparation and characterization of water-soluble chitin and chitosan derivatives. Carbohydr Polym 1998; 36(1): 49-59.
[http://dx.doi.org/10.1016/S0144-8617(97)00235-X]

[67] Muslin T, Morimoto M, Saimoto H, Okamoto Y, Minami S, Shigemasa Y. Synthesis and bioactivities of poly(ethylene glycol)-chitosan hybrids. Carbohydr Polym 2001; 46(4): 323-30.
[http://dx.doi.org/10.1016/S0144-8617(00)00331-3]

[68] Kiuchi H, Kai W, Inoue Y. Preparation and characterization of Poly(ethylene glycol) crosslinked chitosan films. J Appl Polym Sci 2008; 107(6): 3823-30.
[http://dx.doi.org/10.1002/app.27546]

[69] Jeong YI, Kim DG, Jang MK, Nah JW. Preparation and spectroscopic characterization of methoxy poly(ethylene glycol)-grafted water-soluble chitosan. Carbohydr Res 2008; 343(2): 282-9.
[http://dx.doi.org/10.1016/j.carres.2007.10.025] [PMID: 18035341]

[70] Cai G, Jiang H, Tu K, Wang L, Zhu K. A facile route for regioselective conjugation of organo-soluble polymers onto chitosan. Macromol Biosci 2009; 9(3): 256-61.
[http://dx.doi.org/10.1002/mabi.200800153] [PMID: 18855945]

[71] Wang H, Zhao P, Liang X, *et al.* Folate-PEG coated cationic modified chitosancholesterol liposomes for tumor-targeted drug delivery. Biomaterials 2010; 31(14): 4129-38.
[http://dx.doi.org/10.1016/j.biomaterials.2010.01.089] [PMID: 20163853]

[72] Peng H, Xiong H, Li J, Chen L, Zhao Q. Methoxy poly(ethylene glycol)-grafted-chitosan based microcapsules: Synthesis, characterization and properties as potential hydrophilic wall material for stabilization and controlled release of algal oil. J Food Eng 2010; 101(1): 113-9.
[http://dx.doi.org/10.1016/j.jfoodeng.2010.06.019]

[73] Raviña M, Cubillo E, Olmeda D, *et al.* Hyaluronic acid/chitosan-g-poly(ethylene glycol) nanoparticles for gene therapy: an application for pDNA and siRNA delivery. Pharm Res 2010; 27(12): 2544-55.
[http://dx.doi.org/10.1007/s11095-010-0263-y] [PMID: 20857179]

[74] Saito H, Wu X, Harris JM, Hoffman AS. Graft copolymers of poly(ethylene glycol) (PEG) and chitosan. Macromol Rapid Commun 1997; 18(7): 547-50.
[http://dx.doi.org/10.1002/marc.1997.030180703]

[75] Jang M, Nah J. Characterization and modification of low molecular water-soluble chitosan for pharmaceutical application. Bull Korean Chem Soc 2003; 24(9): 1303-7.
[http://dx.doi.org/10.5012/bkcs.2003.24.9.1303]

[76]   Jiang X, Dai H, Leong KW, Goh SH, Mao HQ, Yang YY. Chitosan-g-PEG/DNA complexes deliver gene to the rat liver *via* intrabiliary and intraportal infusions. J Gene Med 2006; 8(4): 477-87.
       [http://dx.doi.org/10.1002/jgm.868] [PMID: 16389625]

[77]   Ouchi T, Nishizawa H, Ohya Y. Aggregation phenomenon of PEG-grafted chitosan in aqueous solution. Polymer (Guildf) 1998; 39(21): 5171-5.
       [http://dx.doi.org/10.1016/S0032-3861(97)10020-9]

[78]   Casettari L, Vllasaliu D, Mantovani G, Howdle SM, Stolnik S, Illum L. Effect of PEGylation on the toxicity and permeability enhancement of chitosan. Biomacromolecules 2010; 11(11): 2854-65.
       [http://dx.doi.org/10.1021/bm100522c] [PMID: 20873757]

[79]   Jeong YI, Kim SH, Jung TY, *et al.* Polyion complex micelles composed of all-trans retinoic acid and poly (ethylene glycol)-grafted-chitosan. J Pharm Sci 2006; 95(11): 2348-60.
       [http://dx.doi.org/10.1002/jps.20586] [PMID: 16886178]

[80]   Hu Y, Jiang H, Xu C, Wang Y, Zhu K. Preparation and characterization of poly(ethylene glycol)--chitosan with water- and organosolubility. Carbohydr Polym 2005; 61(4): 472-9.
       [http://dx.doi.org/10.1016/j.carbpol.2005.06.022]

[81]   Kulbokaite R. Ciuta G Netopilik M, Makuska R. N-PEG'ylation of chitosan *via* "click chemistry" reactions. React Funct Polym 2009; 69(10): 771-8.
       [http://dx.doi.org/10.1016/j.reactfunctpolym.2009.06.010]

[82]   Gorochovceva N, Makuška R. Synthesis and study of water-soluble chitosan-O-poly(ethylene glycol) graft copolymers. Eur Polym J 2004; 40(4): 685-91.
       [http://dx.doi.org/10.1016/j.eurpolymj.2003.12.005]

[83]   Lin WJ, Chen TD, Liu CW, Chen JL, Chang FH. Synthesis of lactobionic acid-grafted-pegylated chitosan with enhanced HepG2 cells transfection. Carbohydr Polym 2011; 83(2): 898-904.
       [http://dx.doi.org/10.1016/j.carbpol.2010.08.072]

[84]   Makuška R, Gorochovceva N. Regioselective grafting of poly(ethylene glycol) onto chitosan through C-6 position of glucosamine units. Carbohydr Polym 2006; 64(2): 319-27.
       [http://dx.doi.org/10.1016/j.carbpol.2005.12.006]

[85]   Huang M, Liu L, Zhang G, Yuan G, Fang Y. Preparation of chitosan derivative with polyethylene glycol side chains for porous structure without specific processing technique. Int J Biol Macromol 2006; 38(3-5): 191-6.
       [http://dx.doi.org/10.1016/j.ijbiomac.2006.02.008] [PMID: 16533520]

[86]   Fangkangwanwong J, Akashi M, Kida T, Chirachanchai S. One-pot synthesis in aqueous system for water-soluble chitosan-graft-poly(ethylene glycol) methyl ether. Biopolymers 2006; 82(6): 580-6.
       [http://dx.doi.org/10.1002/bip.20511] [PMID: 16552764]

[87]   Opanasopit P, Ngawhirunpat T, Rojanarata T, Choochottiros C, Chirachanchai S. N-phthaloylchitosa--g-mPEG design for all-trans retinoic acid-loaded polymeric micelles. Eur J Pharm Sci 2007; 30(5): 424-31.
       [http://dx.doi.org/10.1016/j.ejps.2007.01.002] [PMID: 17307343]

[88]   Corbet C, Ragelle H, Pourcelle V, *et al.* Delivery of siRNA targeting tumor metabolism using non-covalent PEGylated chitosan nanoparticles: Identification of an optimal combination of ligand

structure, linker and grafting method. J Control Release 2016; 223: 53-63.
[http://dx.doi.org/10.1016/j.jconrel.2015.12.020] [PMID: 26699426]

[89] Lukyanov AN, Sawant RM, Hartner WC, Torchilin VP. PEGylated dextran as long-circulating pharmaceutical carrier. J Biomater Sci Polym Ed 2004; 15(5): 621-30.
[http://dx.doi.org/10.1163/156856204323046889] [PMID: 15264663]

[90] Hosseinkhani H, Azzam T, Tabata Y, Domb AJ. Dextran-spermine polycation: an efficient nonviral vector for *in vitro* and *in vivo* gene transfection. Gene Ther 2004; 11(2): 194-203.
[http://dx.doi.org/10.1038/sj.gt.3302159] [PMID: 14712304]

[91] Amini R, Jalilian FA, Abdullah S, *et al.* Dynamics of PEGylated-dextran-spermine nanoparticles for gene delivery to leukemic cells. Appl Biochem Biotechnol 2013; 170(4): 841-53.
[http://dx.doi.org/10.1007/s12010-013-0224-0] [PMID: 23615733]

[92] Suarez Hernandez O, Soliman GM, Winnik FM. Synthesis, reactivity, and pH-responsive assembly of new double hydrophilic block copolymers of carboxymethyldextran and poly(ethylene glycol). Polymer (Guildf) 2007; 48(4): 921-30.
[http://dx.doi.org/10.1016/j.polymer.2006.12.036]

[93] Soliman GM, Winnik FM. Enhancement of hydrophilic drug loading and release characteristics through micellization with new carboxymethyldextran-PEG block copolymers of tunable charge density. Int J Pharm 2008; 356(1-2): 248-58.
[http://dx.doi.org/10.1016/j.ijpharm.2007.12.029] [PMID: 18242897]

[94] Soliman GM, Choi AO, Maysinger D, Winnik FM. Minocycline block copolymer micelles and their anti-inflammatory effects on microglia. Macromol Biosci 2010; 10(3): 278-88.
[http://dx.doi.org/10.1002/mabi.200900259] [PMID: 19937662]

[95] Soliman GM, Szychowski J, Hanessian S, Winnik FM. Robust polymeric nanoparticles for the delivery of aminoglycoside antibiotics using carboxymethyldextran-*b*-poly(ethyleneglycols) lightly grafted with n-dodecyl groups. Soft Matter 2010; 6: 4504-14.
[http://dx.doi.org/10.1039/c0sm00316f]

[96] Karinaga R, Koumoto K, Mizu M, Anada T, Shinkai S, Sakurai K. PEG-appended β-(1>3)-D-glucan schizophyllan to deliver antisense-oligonucleotides with avoiding lysosomal degradation. Biomaterials 2005; 26(23): 4866-73.
[http://dx.doi.org/10.1016/j.biomaterials.2004.11.031] [PMID: 15763266]

[97] Karinaga R, Anada T, Minari J, *et al.* Galactose-PEG dual conjugation of beta-(1>3)-D-glucan schizophyllan for antisense oligonucleotides delivery to enhance the cellular uptake. Biomaterials 2006; 27(8): 1626-35.
[http://dx.doi.org/10.1016/j.biomaterials.2005.08.023] [PMID: 16174528]

[98] Lehtovaara.C, Verma MS, Gu FX. Synthesis of curdlan-*graft*-poly(ethylene glycol) and formulation of doxorubicin-loaded core-shell nanoparticles. J Bioact Compat Polym 2011; 27(1): 3-17.

[99] Altangerel A, Cai J, Liu L, Wu Y, Baigude H, Han J. PEGylation of 6-amino-6-deoxy-curdlan for efficient *in vivo* siRNA delivery. Carbohydr Polym 2016; 141: 92-8.
[http://dx.doi.org/10.1016/j.carbpol.2015.12.077] [PMID: 26877000]

[100] Wakebayashi D, Nishiyama N, Yamasaki Y, *et al.* Lactose-conjugated polyion complex micelles

incorporating plasmid DNA as a targetable gene vector system: their preparation and gene transfecting efficiency against cultured HepG2 cells. J Control Release 2004; 95(3): 653-64.
[http://dx.doi.org/10.1016/j.jconrel.2004.01.003] [PMID: 15023474]

[101] Takae S, Akiyama Y, Otsuka H, Nakamura T, Nagasaki Y, Kataoka K. Ligand density effect on biorecognition by PEGylated gold nanoparticles: regulated interaction of RCA120 lectin with lactose installed to the distal end of tethered PEG strands on gold surface. Biomacromolecules 2005; 6(2): 818-24.
[http://dx.doi.org/10.1021/bm049427e] [PMID: 15762646]

[102] Bergen JM, von Recum HA, Goodman TT, Massey AP, Pun SH. Gold nanoparticles as a versatile platform for optimizing physicochemical parameters for targeted drug delivery. Macromol Biosci 2006; 6(7): 506-16.
[http://dx.doi.org/10.1002/mabi.200600075] [PMID: 16921538]

[103] Sagara K, Kim SW. A new synthesis of galactose-poly(ethylene glycol)-polyethylenimine for gene delivery to hepatocytes. J Control Release 2002; 79(1-3): 271-81.
[http://dx.doi.org/10.1016/S0168-3659(01)00555-7] [PMID: 11853937]

[104] Chen J, Gao X, Hu K, *et al.* Galactose-poly(ethylene glycol)-polyethylenimine for improved lung gene transfer. Biochem Biophys Res Commun 2008; 375(3): 378-83.
[http://dx.doi.org/10.1016/j.bbrc.2008.08.006] [PMID: 18694731]

[105] Oishi M, Nagasaki Y, Nishiyama N, *et al.* Enhanced growth inhibition of hepatic multicellular tumor spheroids by lactosylated poly(ethylene glycol)-siRNA conjugate formulated in PEGylated polyplexes. ChemMedChem 2007; 2(9): 1290-7.
[http://dx.doi.org/10.1002/cmdc.200700076] [PMID: 17546711]

[106] Wang YC, Wu J, Li Y, Du JZ, Yuan YY, Wang J. Engineering nanoscopic hydrogels *via* photo-crosslinking salt-induced polymer assembly for targeted drug delivery. Chem Commun (Camb) 2010; 46(20): 3520-2.
[http://dx.doi.org/10.1039/c002620d] [PMID: 20379597]

[107] Zhu CD, Zheng Q, Wang LX, *et al.* Synthesis of novel galactose functionalized gold nanoparticles and its radiosensitizing mechanism. J Nanobiotechnology 2015; 13: 67-77.
[http://dx.doi.org/10.1186/s12951-015-0129-x] [PMID: 26452535]

[108] Simpson J, Craig D, Faulds K, Graham D. Mixed-monolayer glyconanoparticles for the detection of cholera toxin by surface enhanced Raman spectroscopy. Nanoscale Horiz 2016; 1: 60-3.
[http://dx.doi.org/10.1039/C5NH00036J]

[109] Zhang H, Xiao Y, Cui S, *et al.* Novel galactosylated poly(ethylene glycol)-cholesterol for liposomes as a drug carrier for hepatocyte-targeting. J Nanosci Nanotechnol 2015; 15(6): 4058-69.
[http://dx.doi.org/10.1166/jnn.2015.9707] [PMID: 26369013]

[110] She W, Pan D, Luo K, *et al.* PEGylated dendrimer-doxorubicin conjugates as pH-sensitive drug delivery systems: Synthesis and *in vitro* characterization. J Biomed Nanotechnol 2015; 11(6): 964-78.
[http://dx.doi.org/10.1166/jbn.2015.1865] [PMID: 26353586]

[111] Pontow SE, Kery V, Stahl PD. Mannose receptor. Int Rev Cytol 1992; 137B: 221-44.
[PMID: 1478821]

[112] Zuckerman JE, Gale A, Wu P, Ma R, Davis ME. siRNA delivery to the glomerular mesangium using polycationic cyclodextrin nanoparticles containing siRNA. Nucleic Acid Ther 2015; 25(2): 53-64.
[http://dx.doi.org/10.1089/nat.2014.0505] [PMID: 25734248]

[113] Un K, Kawakami S, Suzuki R, Maruyama K, Yamashita F, Hashida M. Development of an ultrasound-responsive and mannose-modified gene carrier for DNA vaccine therapy. Biomaterials 2010; 31(30): 7813-26.
[http://dx.doi.org/10.1016/j.biomaterials.2010.06.058] [PMID: 20656348]

[114] Yoshida M, Kawakami S, Kono Y, *et al.* Enhancement of the anti-tumor effect of DNA vaccination using an ultrasound-responsive mannose-modified gene carrier in combination with doxorubicin-encapsulated PEGylated liposomes. Int J Pharm 2014; 475(1-2): 401-7.
[http://dx.doi.org/10.1016/j.ijpharm.2014.09.005] [PMID: 25218184]

CHAPTER 4

Recent Developments in Hyaluronic Acid-Based Nanomedicine

Anna Mero[1], Antonella Grigoletto[1], Gabriele Martinez[2] and **Gianfranco Pasut[1,2,*]**

[1] *Department of Pharmaceutical and Pharmacological Sciences, University of Padova, Via F. Marzolo 5, 35131 Padova, Italy*

[2] *Veneto Institute of Oncology IOV-IRCCS, Padova, Italy*

Abstract: Originally considered simply another naturally occurring component of a number of tissues, breakthrough findings later disclosed hyaluronan's (HA) extraordinary biological properties and paved the way for a new era of HA-associated medical applications. HA is, in fact, recognized by many cellular receptors, it can mediate cell migration, proliferation, cell-cell aggregation, it has been shown to promote angiogenesis, *and the list goes on*. This plethora of activities initially moved attention away from a polymer that was considered a simple carrier of biomolecules and towards its potential use in both treating many diseases and conditions and being involved in drug delivery. Given these premises, medical applications exploiting HA's different roles have been developed. The focus of this chapter is directed towards the chemical conjugation of HA with small drugs, peptides and proteins. Reviewing the vast body of literature dedicated to this field, an extraordinary range of applications will be outlined. Although HA cannot be considered a polymer that is appropriate for all uses, conjugates specifically designed to exploit some of its biological properties and for numerous specific applications will no doubt enjoy an advantage over other polymeric conjugates.

Keywords: Anti-TNF-α, Anticancer drug, Anticancer therapy, CD44, Doxorubicin, Drug conjugates, Drug delivery, Human growth hormone, Hyaluronan, Hyaluronic acid, Insulin, Interferon alpha, Osteoarthritis, Paclitaxel,

---

* **Corresponding author Gianfranco Pasut:** Department of Pharmaceutical and Pharmacological Sciences, University of Padova, *Via* F. Marzolo 5, 35131 Padova, Italy; Tel: 617-209-5850; Fax: 617-583-1998; E-mail: gianfranco.pasut@unipd.it

Polymer conjugation, Polysaccharide, Protein conjugation, Protein delivery, RHAMM, RNase A, SN38.

INTRODUCTION

The term "polymer therapeutics" was coined by Ruth Duncan in the 1990s to describe families of polymer constructs to which a drug is covalently bound [1]. This class of substances, which falls under the definition of "new chemical entities", can be distinguished from those drug delivery systems that "per se" only entrap a free drug. Polymer-drug conjugates, polymer-protein/peptide conjugates, polymer-oligonucleotide conjugates, polymeric drugs and block-copolymer-drug micelles all represent sub-classes of polymer therapeutics. Already in 1975, Helmut Ringsdorf introduced the concept of polymer-drug conjugates and the idea of exploiting polymers to increase the solubilization of hydrophobic drugs, to protect drugs from degradation, to prolong their pharmacokinetic profile, and to achieve controlled delivery to cells, tissues and/or organs [2]. Among the several polymers that have been extensively studied, polyethylene glycol (PEG) has proved to be one of the most successful in view of its wide applicability. It has, in fact, already yielded numerous products widely and commonly used in clinical practice such as conjugates with proteins, peptides and oligonucleotides or PEG-modified liposomes [3]. PEG has made a significant contribution to polymer therapeutics, and several new polymers and their derivatives are presently being investigated in the context of preclinical and clinical trials as interest in the development of biodegradable polymers continues to grow. Just a few examples are polyglutamic acid [4], dextrin [5], hyaluronic acid [6], hydroxyethyl starch [7] and polysialic acid [8].

This chapter, which focuses on hyaluronic acid and its conjugates, aims to highlight recent advances in the study of a polymer that can be the carrier for drugs, proteins, and peptides.

General Introduction of Hyaluronic Acid

Hyaluronic acid hyaloid (vitreous) + uronic acid (HA) was isolated for the first time in 1934 from the vitreous humor of cows' eyes by Karl Meyer and John

Palmer [9]. The naturally occurring linear polysaccharide has repeating units of D-glucuronic acid and *N*-acetyl-D-glucosamine disaccharide (Fig. **1**).

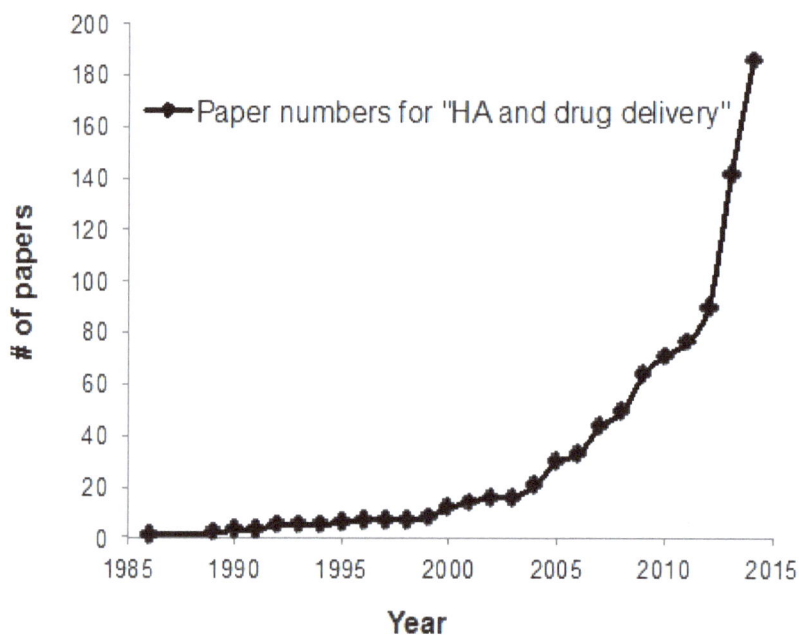

Fig. (1). Publications and citations concerning 'Hyaluronic acid AND drug delivery'. Source: PubMed.

The pK_a of HA's carboxyl groups is therefore 3–4; at pH = 7, the carboxyl groups are predominantly ionized and the polymer is a polyanion that has associated cations (the counterions) and thus referred to as hyaluronan. Found in nature in a wide range of molecular weights (from 20kDa to 4000kDa), HA is a highly hydrophilic polymer that can absorb great quantities of water and expand up to 1000 times its solid volume, forming a loose hydrated network [10].

Hyaluronic acid is present in all vertebrates and in several districts of the human body. It is estimated that the adult human body contains about 12-15 g of hyaluronan, almost half of which is located in the skin [11]. It is also found in the synovial fluid, in the vitreous body of the eye, and it plays an important role in a variety of biological processes: it helps to maintain the viscoelasticity of liquid connective tissues and is involved in controlling tissue hydration, water transport, the organization of the extracellular matrix (ECM) and in tissue repair.

Interestingly, it is also abundantly found in malignancies as well as in inflamed tissues [12].

While it was originally extracted from rooster combs or umbilical cords, hyaluronan is mainly produced for clinical purposes *via* recombinant streptococcal fermentation. As it has an identical chemical structure in all living organisms, it is possible to obtain well-tolerated molecules from various biological sources [13].

HA is implicated in key physiological functions in living organisms including intracellular signaling through specific binding and interaction with hyaladherins that are HA-binding proteins [14], such as the CD44, RHAMM, LYVE-1, IVd4 and LEC receptors [15]. Among these, CD44, one of the cluster of common leukocyte antigens (Clusters of Differentiation, CD), is the most studied HA receptor [16]. CD44 is known to have a variety of significant biological roles including cell–cell aggregation, pericellular matrix retention, matrix-cell and cell-matrix signaling, receptor-mediated internalization/degradation of hyaluronan and cell migration. It also mediates cell migration during morphogenesis, angiogenesis, and tumor invasion and metastasis. It is noteworthy that CD44 receptors are overexpressed in a number of solid cancers. CD44 not only organizes matrix signaling related to cell survival and death, but also mediates the adhesion and rolling of lymphocytes [14]. Unlike other hyaladherins, the HA-mediated motility receptor (RHAMM), also known as CD168, is localized in the cytoplasm and nucleus and is transiently expressed on the surface of activated leukocytes and fibroblasts. RHAMM is known to mediate cell migration and proliferation [17]. Circulating HA in the blood stream is mainly catabolized in liver sinusoidal endothelial cells after internalization by the HA receptor for the endocytosis (HARE) receptor [18].

Due to its versatile biological properties, its biocompatibility and biodegradability, as well as its viscoelastic properties, HA of high, moderate, or low molar mass has found numerous applications in medicine and in cosmetic preparations [19]. Interestingly, HA can be easily modified to tune its physical and biological properties such as solubility, rate of degradation, viscosity and amphiphilicity. Many HA derivatives have been synthesized and are employed in tissue and wound healing and engineering applications; they are used as viscosupplementation agents in joint diseases, as dermal fillings in cosmetic surgery, and as carriers for drug delivery and applications [20 - 22].

HA is considered a particularly promising protein carrier because in addition to the other advantages described above, it has multiple functional groups (hydroxyl and carboxylic acid) along its backbone that can be used for drug conjugation that can generate conjugates with a high drug loading capacity. Its selective interaction with receptors such as CD44 or HARE can also permit an increased accumulation in some specific tissues or organs. HA has been widely investigated as a carrier for receptor-mediated targeting of anticancer drugs and for the delivery of protein, peptide, and nucleotide therapeutics and imaging agents.

HA-Drug Conjugates

HA conjugation to small drugs could be a convenient way to overcome problems such as low water solubility, a short t½, and low specificity. As far as antitumor therapy is concerned, HA's specific interaction with CD44 receptors enables improved drug distribution in tumor tissues, and, in particular, in cancer stem or circulating cells in which CD44 receptors are overexpressed [23]. The growing interest in HA's application in drug delivery applications is reflected in the number of papers and related citations (Fig. **1**) that have appeared in the literature and of conjugates being tested in preclinical or clinical studies (Table **1**).

Table 1. HA-drug-conjugates that have been tested for anticancer activities in preclinical/clinical studies.

DRUG	HA-MW (kDa)	Administration Route	Disease	Tumor Model	Ref.
Aminomethylene diphosphonate	13000			HCT-116 (CD44++)	[61]
Butyric acid	85	*i.t., s.c., i.p.*	Lewis lung, melanoma	NCI-H460, B16F10, Mca, LL3	[62] [63]
Butyric acid, Retinoic acid	85	*i.p.*	Leukemia	P388	[64]
Paclitaxel	200	*i.p., intravesical*	Ovarian, bladder cancer	OVCAR-3, SKOV-3, phase II	[65] [32] [66]

(Table 1) contd.....

DRUG	HA-MW (kDa)	Administration Route	Disease	Tumor Model	Ref.
Paclitaxel	40	*i.p., mtn*	Ovarian cancer	NMP-1, SKOV3ip, Heya8-MDR	[67] [68]
Paclitaxel	40	*i.v.*	SCCHN	OSC-19, HN5	[69]
Paclitaxel	5	*i.v.*	Brain metastasis, breast cancer	231Br	[70]
SN38	200	*i.p.*	Peritoneal cancer	HT-29, MKN-45, OE-21, DHD/K21/Trb	[36] [71]
Doxorubicin	35	*s.c.*	Breast cancer	MDA-MB-468LN	[37]
Doxorubicin	150		Colon cancer	Different cell lines HCT116	[72]
Cisplatin	35	*s.c.*	Breast cancer	MCF-7, MDA-MB231	[41]
Cisplatin	35	*i.t.*	SCCHN	MDA1986	[73]
Cisplatin	35	*i.t.*	Melanoma	A-2058	[44]
Cisplatin	35	*i.t.*	Sarcoma (dog)		[74]
Cisplatin / Doxorubicin	35	*i.t.*	Breast cancer	MDA-MB-468LN	[75]
Cisplatin prodrug	11	*i.v.*	Melanoma	B16-F10	[76]

Abbreviations: i.t., intra-tumor; s.c., subcutaneous; i.p., intraperitoneal; mtn, metronomic therapy; i.v., intravenous.

Carboxylic or hydroxyl groups of HA can be activated directly or through linkers coupled to the drug with a suitable reactive group. The reactions are usually performed in organic conditions, especially in dimethyl sulfoxide (DMSO), using t-Butyl ammonium salt of HA or, whenever possible, in acidic conditions in order to increase the solubility of the polymer. If reactions are performed in water, the pH value should be frequently monitored to prevent polymer hydrolysis. HA is more stable at neutral pH values, and it is more prone to degradation in acidic rather than in basic conditions. High temperatures (>60°C) also induce HA degradation [24].

It is essential to determine the degree of substitution (DS) in HA conjugation in order to maintain the overall polymer charge balance and to exploit some properties, especially in those cases in which the carboxylate group is the site exploited for conjugation. The DS is the ratio between the number of substituted reactive groups and the number of polysaccharides containing repeating disaccharide units (Fig. **2A**). It has been found that a DS ratio above 25% significantly decreases HA's ability to target CD44 receptors.

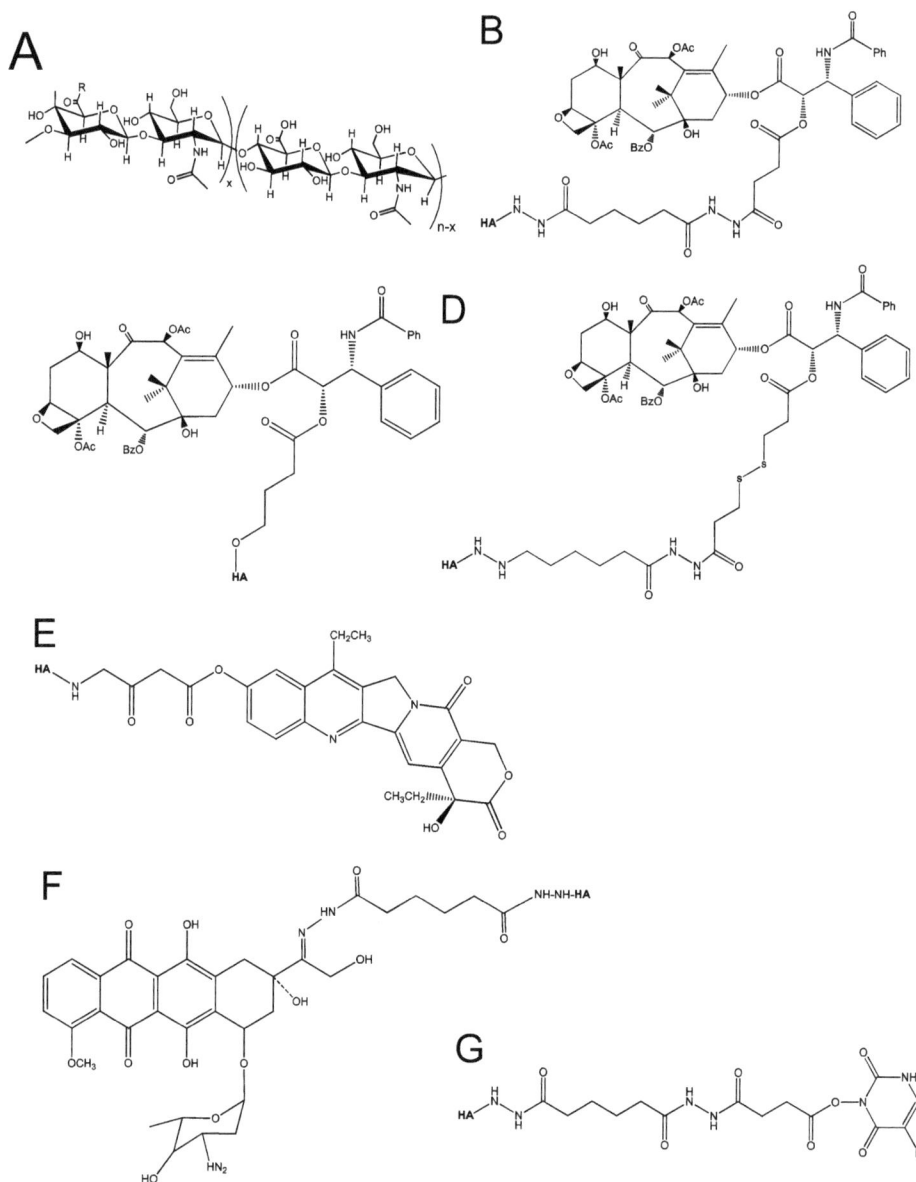

Fig. (2). The chemical structure of some drug conjugates described in this text. (**A**) General structure of HA conjugate with the drug R linked, directly or by means of a spacer, to the HA carboxylic group, (**B**) HA-ADH-PTX, (**C**) HA-butyric acid-PTX (ONCOFID™-P), (**D**) HA-SS-PTX, (**E**) HA-butyric acid-SN38 (ONCOFID™-S), (**F**) doxorubicin-ADH-HA, (**G**) HA-ADH-5FU.

A conjugate of HA with butyrate, a histone deacetylase inhibitor, has been designed to improve cell growth inhibition and apoptosis. The lead compound was a HA of 85 kDa with 4% w/w of butyrate. The conjugate was able to inhibit hepatocellular carcinoma cell growth because of its ability to increase the expression of some cell-cycle related proteins, such as $p21^{wafl}$ and $p27^{kip1}$, and to reduce that of others, including cyclin D1. The effects were further improved by selective targeting of the conjugate to the tumor thanks to an overexpression of CD44 receptors in cancer cells [25].

Paclitaxel (PTX) is an antineoplastic drug that is frequently used to treat solid neoplasms such as breast, lung, prostate and ovarian cancer [26]. PTX suffers from poor water solubility, and its use is characterized by many side effects, also linked to low tolerability to Cremophor, the excipient used in standard formulas to allow PTX solubilisation. Investigators have been attempting since the nineties to conjugate PTX with hydrophilic polymers and HA, among others. HA conjugation of PTX has been investigated using several different strategies and by testing a variety of spacers. Luo and Prestwich engineered a new HA derivative, HA-adipic dihydrazide (HA-ADH), which was conjugated with a N-hydroxysuccinimide (NHS) activated PTX-succinic acid conjugate. Those authors demonstrated that the HA-ADH-PTX conjugate targets CD44 receptors and that it is internalized inside the cells by receptor-mediated endocytosis. The conjugate presents a labile ester bond, between PTX and succinic acid that allowed drug release by intracellular enzymatic hydrolysis (Fig. **2B**). Because of its specific internalization mechanism, it can direct its cytotoxic effect especially towards tumor cells that overexpress CD44s receptors [27].

Tabrizian and co-workers developed a HA-PTX conjugate by coupling PTX-succinic acid-NHS and combined the prodrug approach with the layer-by-layer (LbL) technique. A hyaluronan ester prodrug of paclitaxel was used as the polyanion to construct a polyelectrolyte multilayer (PEM) with chitosan (CH), which is a polyamine. The HA-PTX conjugate was obtained by coupling PTX-succinic acid-NHS with a HA-ethylendiamine (HA-EDA) derivative. The new conjugate was assembled in a PEM construct with CH to control degradation and drug release. HA played the dual role of a structural element of the PEM and a macromolecular carrier of the drug. The efficacy of these paclitaxel-loaded PEMs was investigated in a cell viability assay using J774 macrophages. The cells cultured onto paclitaxel loaded PEMs presented a 95% reduction in viability with respect to the control and to the unmodified multilayer (100.2% cell survival) [28].

In another approach, the hydroxyl group of PTX was first esterified with 4-bromobutyric acid; the derivative was then conjugated to HA of 200kDa through the formation of a second ester bond [29, 30]. The conjugate, called ONCOFID™-P (Fig. 2C), showed superior properties such as high drug loading and solubility with respect to the HA-PTX conjugates in which the PTX was directly coupled to the carboxylic groups of HA. PTX was promptly released from ONCOFID™-P after cell internalization and the conjugate presented higher activity with respect to free PTX [31]. ONCOFID™-P has also showed promising results in a model of ovarian cancer in mice [32]. To date, ONCOFID™-P is undergoing phase II clinical studies [31, 33].

Yin and coworkers designed a new HA conjugate of PTX using a disulfide spacer (HA-SS-PTX; Fig. 2D). A disulfide bridge was introduced to allow drug release in tumor cells by disulfide reduction in response to higher glutathione concentrations. It was reported that intracellular uptake of the conjugate into MCF-7 cells was regulated by CD44-caveolae-mediated endocytosis. Among the different sizes of HA used in the study (MW 9.5, 35 and 770kDa), HA9.5-S-PTX showed the best results with improved tumor growth inhibition, drug loading, cellular internalization, and tumor targeting capability with respect to a PEGylated PTX obtained with the same disulfide spacer [34].

Camptothecin-11 (CPT-11) is a potent antineoplastic drug that inhibits topoisomerase but is nevertheless characterized by poor solubility and stability. Indeed, the drug contains a lactone ring that can be hydrolyzed to the open inactive form in aqueous environments. Polymer conjugation of CPT has been studied to overcome these problems. There are many examples of CPT-11 prodrugs, also prepared with CTP-11 analogs such as topotecan and irinotecan.

When the researchers of the Fidia Farmaceutici Company coupled an active metabolite of irinotecan 7-ethyl-10-hydroxycamptotechi (SN38) to HA, they obtained a conjugate called ONCOFID™-S (Fig. **2E**). The derivative was specifically designed for the loco-regional intraperitoneal treatment of peritoneal carcinomatosis [35]. The chemistry of the conjugation was the same one utilized above for the synthesis of ONCOFID™-P that is based on a double ester spacer. ONCOFID™ S exerts a strong *in vitro* anti-proliferative activity on CD44 over-expressing rat DHD/K12/trb colon adenocarcinoma cells, as well as of gastric, breast, esophageal, ovarian and lung human cancer cells, activities that were superior to those for SN-38 alone. ONCOFID™-P was also tested *in vivo* in two preclinical models of colorectal cancer (CRC) in BDIX rats; it was demonstrated that once administered intraperitoneally, it was able to dramatically reduce all parameters indicative of a poor prognosis in peritoneal metastatization of CRC without any myelotoxicity or mesothelial inflammation [33, 36].

Although presenting renal, hepatic and cardiac toxicities, Doxorubicin (DOX) is a very effective chemotherapeutic. In order to improve its effectiveness and to reduce toxicity, DOX has been conjugated with HA 35kDa through an ADH linker to obtain a hydrazone acid labile linkage [37] (Fig. **2F**). The drug loading ranged from 5 to 15% (w/w). Cardiac toxicity, pharmacokinetic and tumoral activity were evaluated by *in vivo* models. A peritumoral injection led to delayed tumor progression by approximately 10 weeks, and it increased the animals' survival with respect to intravenous DOX treatment.

5-Fluorouracil (5FU) has also been conjugated with HA using ADH and succinic acid spacers (Fig. **2G**). The polymeric prodrug with a drug loading of 87.67 mg/g presents a high stability in an acidic environment and moderate stability in a HAase solution and plasma. This conjugate has presented intrinsic active tumor targeting of HA and anti-proliferative effects in several cancer cell lines. HA-5FU was evaluated for its anti-tumor activity on A2780 and HepG2. It was found that cytotoxicity depended on drug concentration, incubation time, and the cell type, and that the conjugate was more cytotoxic than the free drug. Pharmacokinetic studies showed an improved AUC and half-life in comparison with the free drug [38].

Methotrexate (MTX) is an antimetabolite and an analogue of folic acid that is used as an antineoplastic drug. Eurand Pharmaceuticals developed a HA-MTX prodrug by introducing at the C6 position a chloride of HA to allow reaction with MTX. Acetyl or butyryl groups were introduced to substitute the unreacted chloride, thus yielding mixed esters. Not involving the carboxylic groups of HA has the advantage of increasing the conjugate's water solubility. The formation of a specific ester allows a well-defined and reproducible release profile. This conjugate presented a drug loading between 10-18% [39]. In the HA-MTX conjugate designed by Homma and co-workers, the drug was linked to a lysosomal enzyme sensitive peptide linker (Asn-Phe-Phe) and was then conjugated to HA through a spacer that avoided the potential steric entanglement of the polymer towards the proteases intended to cleave the peptide. After intra-articular injection, the conjugate with the Asn-Phe-Phe linker was better able to reduce knee swelling in arthritic rats with respect to free MTX. A similar conjugate prepared with Gly-Phe-Leu-Gly peptide as a linker and with a similar MTX loading inhibited knee swelling only slightly suggesting that the cleavage of the conjugate *in vivo* depends on the peptide sequence [40].

Cisplatin (CDDP) is an anticancer therapy used to treat sarcomas and cancers of the bones, muscles and blood vessels. It presents several dose-dependent side effects such as nephrotoxicity, neurotoxicity and myelosuppression. Several HA conjugates have been designed to increase its tumor delivery and to reduce its toxicity. Forrest's group linked CDDP to carboxyl groups of HA (35kDa) using silver nitrate as the activating agent. The conjugate increased drug concentration in loco-regional nodal tissues with respect to the standard formula and resulted in sustained drug release [41, 42]. *In vivo* studies reported selective anticancer effects [43]. When HA-CDDP was tested in a melanoma cell line A-2058 xenograft model, the conjugate was more effective in inhibiting tumor growth with respect to a free CDDP *i.v.* injection [44].

To reduce Pt(II) toxicity, Ling and coworkers developed a stimuli sensitive Pt(IV) prodrug linked to ethylendiamine (EDA) modified hyaluronic acid to form a tumor-targeting HA-EDA-Pt(IV) conjugate. Preserving the tumor targeting properties, this system showed sustained drug release. After internalization of HA-EDA-Pt(IV), the Pt(IV) was reduced to the active Pt(II) form. Toxicological evaluation demonstrated that toxic side effects were significantly alleviated [45].

HA has also been conjugated with several anti-inflammatory drugs containing a hydroxyl group such as hydrocortisone, prednisone, prednisolone, fluorocortisone, dexamethasone, betamethasone and corticosterone. The hydroxyl group was directly conjugated to HA's carboxylic group by esterification, and conjugates were developed for intra-articular arthritis treatment [46, 47].

HA–Peptide Conjugates

There are a variety of strategies for the bioconjugation of HA to biologics. While the simplest strategy is based on the direct activation of HA carboxyl groups, the procedure should be carefully evaluated, because the risk of cross-linking and loss of activity is high as the peptides, and proteins even more, present several reactive side-chains. Using this strategy, Ferguson conjugated HA to epidermal growth factor (EGF) with a condensating agent [48]. EGF was not active *in vitro* after random modification probably because the random modification of the peptide inhibited receptor binding, thus preventing the proliferative activity of EGF.

We have recently proposed an alternative approach to selectively modify peptides and proteins. The method is based on introducing aldehyde groups by coupling 4-aminobutyraldehyde diethyl acetal spacers to some carboxylic groups of HA. This approach makes it possible to exploit a site-selective conjugation chemistry also known as N-terminal amino conjugation. The low reactivity of the aldehyde group favors coupling at the less basic N-terminal amino group with respect to the epsilon-amino group of lysine when the reaction is performed under acidic conditions that induces strong protonation of the more basic group [49]. Exploiting this strategy we successfully modified salmon calcitonin (sCT) with a HA-aldehyde derivative. sCT is now considered a treatment option for osteoarthritis, a disease characterized by chronic deterioration of the articular cartilage of the joints and the underlying bone. Recent studies have demonstrated that sCT acts by inhibiting both bone turnover and cartilage degradation and by reducing matrix metalloproteinases (MMP) activities [50]. sCT nevertheless shows a rapid clearance from the joint thus preventing prolonged local action and causes undesired systemic effects. In view of the fact that high molecular weight HA is already being used as a viscosupplement in osteoarthritis and has been shown to normalize the levels of MMP-1, -3 and -13, we designed a HA-salmon calcitonin (HA-sCT) conjugate for the local treatment of osteoarthritis [51]. HAylation of sCT was considered a reasonable approach to combine the two synergic actions and to delay the rapid clearance of sCT from the knee articular space after intra-articular injections. The conjugate showed good preservation of sCT activity and prolonged residence time in the joint space, reducing systemic exposure to sCT [51]. A promising chondroprotective effect has also been demonstrated in a rabbit model of osteoarthritis.

An alternative strategy was utilized by Kong and co-workers who selectively modified Exendin, an anti-diabetic peptide for the treatment of type 2 diabetes. After HA was modified with vinyl sulfone, Exendin-4, which was modified at the C-terminus by adding a cysteine, was selectively linked to the polymer with the Michael's addition reaction. The ability of free and HAylated peptide to lower the glucose levels was tested on Type 2 diabetic db/db mice. The blood glucose levels decreased after the injection of the free peptide and started to return to normal levels within a day's time; the conjugated peptide, instead, exerted its effect for as long as three days. The insulinotropic bioactivity of HA–Exendin-4 conjugate was confirmed by insulin immunohistochemical analysis of the pancreas tissues of Type 2 diabetic db/db mice isolated 30 min after sample injection. A denser color, reflecting more insulin secretion, was visible after injection of the HA–Exendin-4 conjugate with respect to the PBS or free peptide, confirming the conjugate's potential for treating Type 2 diabetes with longer intervals between injections (which were administered twice a week instead of every day) [52].

Retinal and choroidal vascular disease and diabetic retinopathy are pathologies that are linked to an overexpression of vascular endothelial growth factor (VEGF) in the retina that leads to angiogenesis and hyperpermeability [53]. Anti-FLT1 peptide is an antagonist peptide of seven amino acids against VEGF receptor 1 (VEGFR1) [20] that specifically binds to VEGFR1 inhibiting the interaction of VEGFR1 with a variety of VEGFR1 ligands, such as VEGFA, VEGFB, and placental growth factor (PlGF). Oh and co-workers proposed conjugating anti-FlLT1 peptide to HA. The peptide was chemically conjugated to HA in anhydrous DMSO using benzotriazol-1-yloxytris (dimethylamino) phosphonium hexafluorophosphate (BOP) as the coupling reagent. It was found, first of all, that HA greatly increased the peptide's solubility in the aqueous medium, making it possible to verify if the conjugate peptide has a high activity *in vitro*. Moreover, thanks to the viscoelastic and mucoadhesive properties of the polymer, the conjugated peptide efficaciously inhibited retinal choroidal neovascularization (CNV) in laser-induced CNV model rats. The retinal vascular permeability and the deformation of the retinal vascular structure were also significantly reduced in diabetic retinopathy model rats after treatment with HA–Anti-FLT1 conjugate. Finally, the conjugate acquired a micelle conformation in aqueous solution with the polymers covering the hydrophobic peptide (Fig. **3**). As confirmed by pharmacokinetic analysis, this stricture preserves the degradation of the peptide increasing the residence time of conjugated peptide to more than two weeks [54].

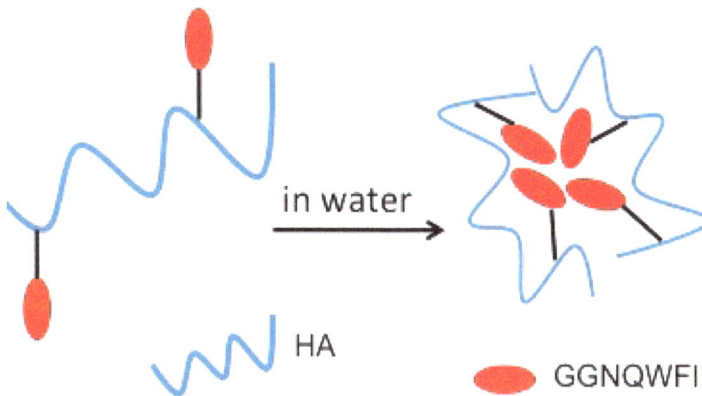

Fig. (3). Schematic representation of HA-Peptide Conjugate. In aqueous solution the conjugate acquires a micelle configuration that allows a prolonged residence time in the bloodstream.

HA-Protein Conjugates

As reported above with regard to peptides, also for proteins the direct activation of HA's carboxylic group and subsequent conjugation to the amino group of lysines would be easy and rapid process but could present the high risk of yielding cross-linked products. The direct activation of HA carboxylic groups in water with EDC and NHS was studied for trypsin [48] based on the idea that the polymeric backbone is degraded *in vivo* by hyaluronidase (HAase) thus unmasking the conjugated protein and recovering its activity. Unlike in the case of HA-EGF, the activity of unmasked trypsin increased up to 145% with respect to the free enzyme after incubation with physiological concentrations of HAase [48]. These contrasting results seem to highlight the difficulties linked to direct HA activation for protein conjugation. Indeed, the approach outlined led to a random modification of the proteins because it involved the free amino groups present on their surface. As lysines are usually fairly abundant in proteins, this strategy may not always be successful because the polymer can be closely linked to a crucial domain for protein activity and therefore become a steric entanglement. A random modification could thus be unreliable especially with regard to proteins that interact with receptors. Due to this low selective chemistry, mixtures of isomers were obtained, prejudicing the purification and characterization steps and the batch-to-batch reproducibility. As the requirements for approval of pharmaceuticals for humans are stringent, only thoroughly characterized conjugates can meet the specifications outlined by regulatory agencies. The attention of researchers is thus mainly directed to conjugation approaches that can yield a single point of coupling in a protein sequence [55].

Yang *et al.* proposed the site-specific conjugation of interferon alpha (IFNα) to a hyaluronan-aldehyde derivative in acidic condition, pH 5.5 [56]. The coupling reaction was directed to the *N*-terminal primary amine group of IFNα to obtain a range of 2–9 protein units linked per polymer chain. The biological activity of HA-IFNα was investigated *in vitro* by testing the anti-proliferative activity of the cytokine on Daudi cells. With respect to IFNα and PEG-Intron, the anti-proliferation of HA-IFNα was lower than that of native IFNα but comparable to that of PEGylated IFNα. Dye labeled HA-IFNα demonstrated a specific delivery of the conjugate to the liver; IFNα was instead distributed and then eliminated by renal clearance. The pharmacokinetic profiles of IFNα and HA-IFNα showed that a single administration of HA-IFNα resulted in the circulation of IFNα for almost 5 days depending on the degree of modification of HA; native IFNα instead underwent a rapid clearance within 24 h. Finally, *in vivo* antiviral activity was evaluated based on the expression in the liver tissue of the antiviral protein OAS 1 induced by IFNα. Slightly modified HA-IFNα showed an OAS 1 expression level higher than the highly modified one and PEG-Intron, perhaps given the better delivery of slightly modified HA derivatives to the liver. In this work, HA-aldehyde was prepared by mild oxidation of the polymeric backbone with sodium periodate. This procedure generated two close aldehyde groups for each saccharide oxidized unit consequent to the oxidation of two vicinal hydroxyl groups and the subsequent ring opening. The proximity of the two aldehyde groups complicates determining which aldehyde is involved in the protein coupling, or, if both are reacting, to what extent. Moreover, the ring opening of a certain number of saccharide units modifies the HA polymeric backbone and this might affect some of its properties. Researchers also conjugated HA to hGH following this same procedure. The conjugate showed activity in fibroblasts by elevating the hGH-receptor mediated signaling pathway. HA-hGH, administered by topical treatment, demonstrated an AUC and a bioavailability 10 and 16 fold higher, respectively, with respect to topically delivered hGH [57].

Friedrich *et al.*, who proposed the HA conjugation of anti-TNF-α for application in wound healing following burn injuries, demonstrated a sustained and local modulation of post-injury inflammatory responses [58]. In particular, they considered the effects on burn progression and on the inflammatory microenvironment of topical application of free anti-TNF-α, (anti-TNF-α)-HA and a mixture of the two. As is known, TNF-α is an upstream regulator of inflammation, having a variety of potent effects and it is present at significant levels in burn wound tissue and in wound fluid. Local and controlled modulation of TNF-α signaling could feasibly result in significant improvements in the healing process. A cytokine-neutralizing anti-TNF-α antibody was randomly conjugated to HA (M_w = 1.6 MDa) after HA's carboxylic groups were directly activated in order to minimize the protein's side effects, increase its effectiveness and mediate inflammation locally. The final conjugate solution was prepared containing 5% of HA, by mixing 3 parts of the conjugate product with 4 parts of 8% HA solution, thus increasing the viscosity. When the binding affinity of the conjugated antibody with respect to free anti-TNF-α was checked, it was found that conjugation to high molecular weight HA derivatives did not compromise the affinity of the mAb for TNF-α. After burn injuries were induced in the rats, a histological analysis and protein extraction was carried out just as were vimentin quantification, macrophage infiltration and evaluation of interleukin-1-beta (IL-1β) levels. At an initial analysis made after 4 days, a layer of dermal tissue following secondary necrosis had grown significantly in anti-TNF-α treatment, but not in the (anti-TNF-α)-HA one. Moreover, the viable tissue around the wound appeared healthiest for the (anti-TNF-α)-HA treatment at day 7 and the areas of multinucleated immune cells packed at the wound edge were less dense in the (anti-TNF-α)-HA treatment with respect to all the other groups, suggesting an attenuation of the acute inflammatory response. The identification of vimentin, which is an intermediate filament found in mesodermal tissue, made it possible to estimate the extent of nonviable or significantly compromised tissue as an index of burn progression. Anti-TNF-α and anti-TNF-α + HA treatments appeared to produce a relative decrease in nonviable tissue compared to saline, while (anti-TNF-α)-HA treatment was associated to the greatest decrease which was significantly lower than that of all other treatments. Immunostaining for CD68 was adopted for the estimation of the total number of macrophages in the wound bed. Treatment with anti-TNF-α or (anti-TNF-α)-HA significantly affected macrophage counts with respect to saline. Although anti-TNF-α + HA followed this same pattern at later time points, it had no effect on day 1. Treatment with non-conjugated anti-TNF-α appeared to have the greatest effect in reducing macrophage infiltration, while (anti-TNF-α)-HA and anti-TNF-α + HA appeared to have similar effects. Finally, IL-1β concentrations were evaluated, as this cytokine is mainly involved in inflammation and linked to the levels of active TNF-α. The results suggested that sites treated with (anti-TNF-α)-HA showed a real reduction in IL-1β levels at days 4 and 7 with respect to saline.

Overall, (anti-TNF-α)-HA reduced nonviable tissue and related inflammatory markers more effectively than anti-TNF-α, HA or anti-TNF-α + HA, as it was demonstrated that conjugation of anti-TNF-α to high molecular weight HA provided sustained, local modulation of the post-injury inflammatory responses compared to direct administration of non-conjugated antibodies.

The hyaluronan-aldehyde derivative proposed by us, described above for the conjugation of sCT, is a suitable platform for selective *N*-terminal protein conjugation [49]. The advantage of this HA-aldehyde derivative is that the integrity of the polymer backbone is preserved. The acetal spacer was chosen because it ensures high stability of the aldehyde groups during long-term storage as the final HA-acetal derivative can be easily deprotected in acidic conditions to prepare the aldehyde active form when it is required for protein conjugation. It moreover simplifies the characterization of the conjugates in view of a single pathway coupling because it is less reactive than other shorter aliphatic aldehydes, namely propyl or ethyl aldehydes. The potentialities of this HA-acetal were first tested by coupling to model proteins. HAylation of RNase A, trypsin and insulin with the new HA-aldehyde derivative uncovered some important results, such as good retention of enzymatic activity and increased thermal stability of proteins. In particular, for trypsin, the polymer reduced the typical autolysis process. After incubation in physiological conditions at 37 °C for 6 h, the free enzyme had lost more than 90% of activity, whereas for the conjugate, the loss was only of 34%.

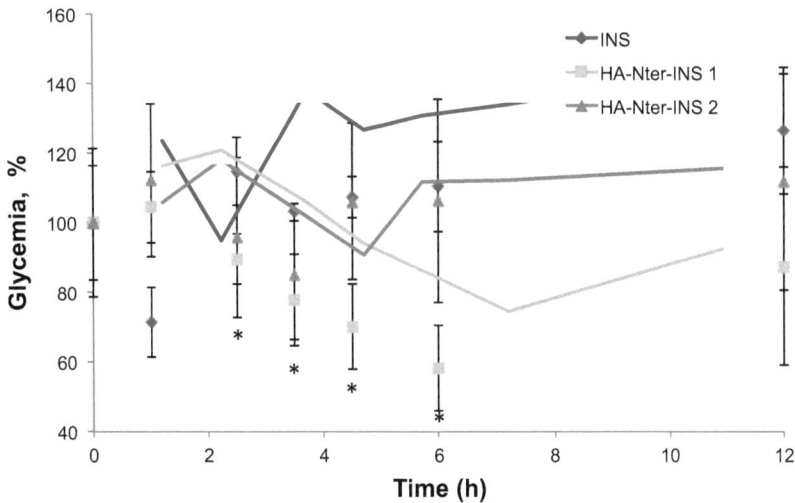

Fig. (4). Pharmacodynamics of HA-INS conjugates. Blood glucose levels in streptozocin-induced diabetic Sprague-Dawley rats after HA-Nter-INS 1, HA- Nter-INS 2, bovine insulin injections. The results are presented as mean ± s.e.m. (n=3-5). Significance: * = p < 0.05 *vs* insulin. Adapted with permission from [49].

Two HA–insulin (HA-INS) conjugates were synthetized with different INS drug payloads. A *in vivo* study on diabetic rats showed that the conjugate with a lower degree of INS (HA-Nter-INS 1) loading had a glucose lowering effect that lasted up to 6 h, while free INS exhausted its action after 1 h (Fig. **4**). Unexpectedly, the conjugate with the higher loading (HA-Nter-INS 2) showed no significant effect on the glucose titer, probably due to the steric entanglement affecting the receptor/protein recognition derived from the higher insulin loading.

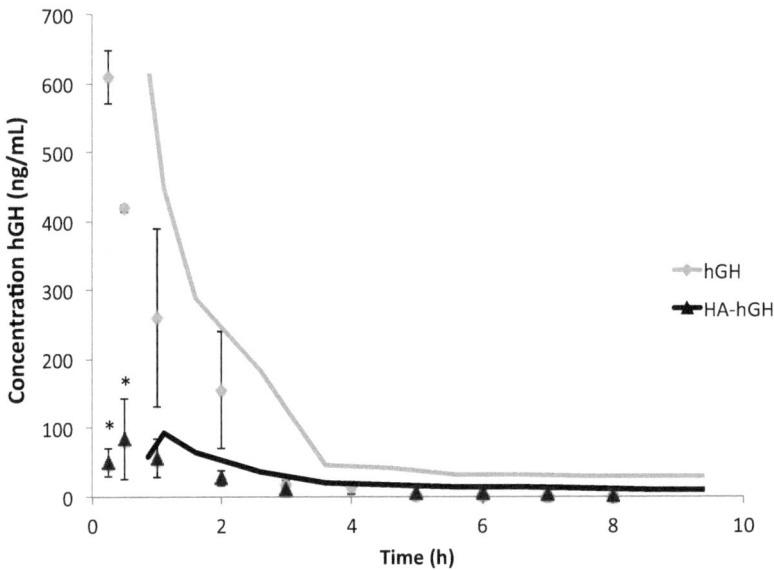

Fig. (5). Blood hGH levels after an i.a. administration of native and HAylated hGH in Sprague-Dawley rats (n = 5 per group). The animals received 100 µg of hGH equivalent. The profiles demonstrated that with respect to the conjugate, hGH cleared quickly from the joint space. Data represents mean ± S.D. Significance: * = p < 0.01 *vs* hGH. Adapted with permission from [59].

Mero *et al.* also described the advantages of HA conjugation with respect to hGH when utilized to treat osteoarthritis (OA) [59]. hGH has anabolic effects on articular joints and probably prompts cartilage growth through local and systemic IGF-1 production by directly stimulating chondrocyte maturation. When this protein is used to treat localized diseases, such as OA, intra-articular (*i.a.*) administration could promote undesired systemic effects due to rapid clearance from the articular space. HA injections are nevertheless mainly used for their viscosupplementation effect on the synovial fluid and for the several pharmacologic activities outlined above. The rationale behind this therapeutic approach resides in the intent of combining hGH's activity with HA's inhibition of MMP-1, -3, and -13 expression. In addition, besides reducing hGH clearance from the joint space (Fig. **5**), HAylation of hGH also slightly increased hGH stability, showing the absence of formation of large aggregates after incubation at temperatures reaching 80°C while hGH underwent aggregation at 70°C. Although *in vitro* studies demonstrated that conjugation reduced hGH activity, *in vivo* the situation might be counterbalanced by the retention of the protein in the articular space which would thus enhance its local effect. Protein bioactivity for HA conjugates can be partially recovered *in vivo* by hyaluronidase that degrades the polymeric backbone, reducing the length of HA chains and consequently their steric entanglement in HA-hGH / hGH receptor recognition.

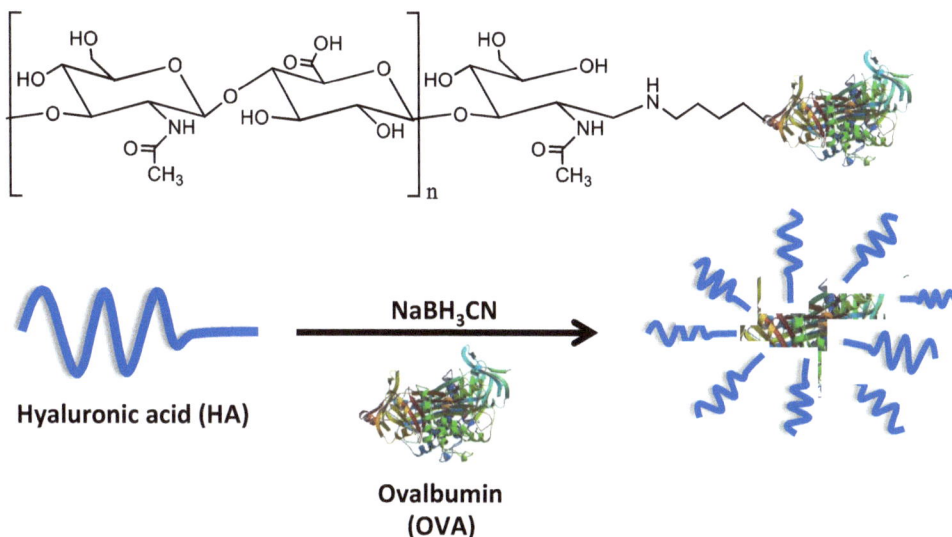

Fig. (6). Synthesis of a polymer conjugate for targeted delivery of foreign antigen (OVA). The HA-OVA conjugate was prepared by reductive amination between the reducing end of HA to the amino group of OVA.

Lee *et al.* investigated the modification of ovalbumin (OVA) with hyaluronic acid in an immunotherapy study in which the synthetized derivative was utilized as a tumor-targetable polymer conjugate with a foreign antigen capable of inducing immunological rejection of tumors [60]. As a model system, the foreign antigen OVA was decorated with HA in order to exploit the polymer's targeting tumor properties and to generate an antitumor effect by provoking immune rejection against tumors presenting foreign antigens. The investigators aimed to deliver non-self foreign antigens into the tumors thus inducing rejection by foreign antigen-reactive cytotoxic T lymphocytes (CTLs). HA was chosen as the targeting moiety because tumor cells over-express HA receptors such as CD44 and RHAMM. The targeted delivery of OVA to tumor cells could facilitate the presentation of protein fragments on the cellular surface, thus allowing antigen-specific CTLs to kill the tumor. The HA-OVA conjugate was prepared by the random conjugation of HA to OVA through reductive amination between the reducing end group of HA and the amino group of OVA in the presence of $NaBH_3CN$ as a reducing agent (Fig. **6**). The resulting conjugate is presented as a single molecule of OVA decorated with nine HA molecules of 35kDa.

The HA shell on the protein surface may increase receptor-mediated endocytosis by cancer cells over-expressing the HA receptors. Furthermore, after endocytosis, OVA release could be enhanced, leading to presentation of the OVA antigen on the cell surface by the MHC class I molecule because HA is readily fragmented in intracellular compartments containing hyaluronidase. To visualize the cellular uptake of the conjugate *in vitro*, HA-OVA and OVA were FITC-labeled and HA-OVA-FITC showed much stronger fluorescence signal at the intracellular level than did OVA-FITC, exhibiting 3.27-fold higher uptake by TC-1 cells. The antigen presentation behavior after treatment with the conjugate was quantitatively evaluated both *in vitro* and *in vivo* using a flow cytometer. *In vitro* HA-OVA conjugate induced much higher OVA peptide presentation with respect to the physical mixture of HA/OVA or OVA alone. Moreover, TC-1 tumor cells treated with OVA, HA/OVA, or HA-OVA were incubated with OVA-specific CTLs and the conjugate resulted in significantly enhanced apoptosis of TC-1 tumor cells by OVA-specific CTLs with respect to native OVA. The *in vivo* biodistribution showed that HA-OVA had the highest tumor target ability after systemic administration into the tumor-bearing mice. OVA antigen presentation *in vivo* was 4.2-fold higher for the TC-1 tumor treated with the HA-OVA with respect to the free OVA treated one. When generation of OVA-specific CD8+ T cells following vaccinia virus encoding recombinant OVA immunization was evaluated, the mice with HA-OVA possessed the largest number of CTLs. Finally, while the tumor volume did not significantly increase when the mice were treated with HA-OVA conjugate, the groups that received OVA and HA/OVA showed rapid increase in tumor volume as a function of time, implying no antitumor activity.

CONCLUSION

This review highlights some of the numerous possible applications as well as the potential uses of HA. Although sharing some of the common problems of natural polymers such as high polydispersity and a certain degree of instability, HA's intrinsic biological properties permit it to play more than just the simple role of an inert polymeric carrier. The polymer is a therapeutic agent and an indicator of disease processes; its high viscoelasticity has been exploited to treat osteoarthritis and it has been found to reduce nerve impulses and sensitivity associated with the activity of proinflammatory mediators and MMP. The promising results that have been reported in the literature until now have confirmed HA's wide applicability in a variety of areas and we foresee even more and better clinical applications when its biological properties have been fully understood and can be exploited from the time a new conjugate is first imagined.

CONFLICT OF INTEREST

The authors confirm that they have no conflict of interest to declare for this publication.

ACKNOWLEDGEMENTS

Ministry of Education University and Research (MIUR; Grant No.60A04-9402/15)

REFERENCES

[1]   Duncan R. The dawning era of polymer therapeutics. Nat Rev Drug Discov 2003; 2(5): 347-60.
        [http://dx.doi.org/10.1038/nrd1088] [PMID: 12750738]

[2]   Duncan R, Vicent MJ. Polymer therapeutics-prospects for 21st century: the end of the beginning. Adv Drug Deliv Rev 2013; 65(1): 60-70.
        [http://dx.doi.org/10.1016/j.addr.2012.08.012] [PMID: 22981753]

[3]   Pasut G, Veronese FM. State of the art in PEGylation: the great versatility achieved after forty years of research. J Control Release 2012; 161(2): 461-72.
        [http://dx.doi.org/10.1016/j.jconrel.2011.10.037] [PMID: 22094104]

[4]   Bajaj I, Singhal R. Poly (glutamic acid)an emerging biopolymer of commercial interest. Bioresour Technol 2011; 102(10): 5551-61.
        [http://dx.doi.org/10.1016/j.biortech.2011.02.047] [PMID: 21377358]

[5]   Ferguson EL, Duncan R. Dextrin-phospholipase A2: synthesis and evaluation as a bioresponsive anticancer conjugate. Biomacromolecules 2009; 10(6): 1358-64.
        [http://dx.doi.org/10.1021/bm8013022] [PMID: 19354276]

[6]   Tripodo G, Trapani A, Torre ML, Giammona G, Trapani G, Mandracchia D. Hyaluronic acid and its

derivatives in drug delivery and imaging: Recent advances and challenges. Eur J Pharm Biopharm 2015; 97(Pt B): 400-16.

[7] Sharma AR, Kundu SK, Nam JS, *et al.* Next generation delivery system for proteins and genes of therapeutic purpose: why and how? Biomed Res Int 2014; 2014: 327950.
[http://dx.doi.org/10.1155/2014/327950]

[8] Gregoriadis G, Jain S, Papaioannou I, Laing P. Improving the therapeutic efficacy of peptides and proteins: a role for polysialic acids. Int J Pharm 2005; 300(1-2): 125-30.
[http://dx.doi.org/10.1016/j.ijpharm.2005.06.007] [PMID: 16046256]

[9] Meyer K, Palmer JW. The polysaccharide of the vitreous humor. J Biol Chem 1934; 107(3): 629-34.

[10] Laurent TC, Fraser JR. Hyaluronan. FASEB J 1992; 6(7): 2397-404.
[PMID: 1563592]

[11] Karbownik MS, Nowak JZ. Hyaluronan: towards novel anti-cancer therapeutics. Pharmacol Rep 2013; 65(5): 1056-74.
[http://dx.doi.org/10.1016/S1734-1140(13)71465-8] [PMID: 24399703]

[12] Cowman MK, Matsuoka S. Experimental approaches to hyaluronan structure. Carbohydr Res 2005; 340(5): 791-809.
[http://dx.doi.org/10.1016/j.carres.2005.01.022] [PMID: 15780246]

[13] Liu L, Liu Y, Li J, Du G, Chen J. Microbial production of hyaluronic acid: current state, challenges, and perspectives. Microb Cell Fact 2011; 10(1): 99.
[http://dx.doi.org/10.1186/1475-2859-10-99] [PMID: 22088095]

[14] Choi KY, Saravanakumar G, Park JH, Park K. Hyaluronic acid-based nanocarriers for intracellular targeting: interfacial interactions with proteins in cancer. Colloids Surf B Biointerfaces 2012; 99: 82-94.
[http://dx.doi.org/10.1016/j.colsurfb.2011.10.029] [PMID: 22079699]

[15] Sherman L, Sleeman J, Herrlich P, Ponta H. Hyaluronate receptors: key players in growth, differentiation, migration and tumor progression. Curr Opin Cell Biol 1994; 6(5): 726-33.
[http://dx.doi.org/10.1016/0955-0674(94)90100-7] [PMID: 7530464]

[16] Marhaba R, Zöller M. CD44 in cancer progression: adhesion, migration and growth regulation. J Mol Histol 2004; 35(3): 211-31.
[http://dx.doi.org/10.1023/B:HIJO.0000032354.94213.69] [PMID: 15339042]

[17] Hardwick C, Hoare K, Owens R, *et al.* Molecular cloning of a novel hyaluronan receptor that mediates tumor cell motility. J Cell Biol 1992; 117(6): 1343-50.
[http://dx.doi.org/10.1083/jcb.117.6.1343] [PMID: 1376732]

[18] Zhou B, Weigel JA, Fauss L, Weigel PH. Identification of the hyaluronan receptor for endocytosis (HARE). J Biol Chem 2000; 275(48): 37733-41.
[http://dx.doi.org/10.1074/jbc.M003030200] [PMID: 10952975]

[19] Arshinoff SA, Jafari M. New classification of ophthalmic viscosurgical devices 2005. J Cataract Refract Surg 2005; 31(11): 2167-71.
[http://dx.doi.org/10.1016/j.jcrs.2005.08.056] [PMID: 16412934]

[20]   Bae DG, Kim TD, Li G, Yoon WH, Chae CB. Anti-flt1 peptide, a vascular endothelial growth factor receptor 1-specific hexapeptide, inhibits tumor growth and metastasis. Clin Cancer Res 2005; 11(7): 2651-61.
[http://dx.doi.org/10.1158/1078-0432.CCR-04-1564] [PMID: 15814646]

[21]   Davidson JM, Nanney LB, Broadley KN, *et al.* Hyaluronate derivatives and their application to wound healing: preliminary observations. Clin Mater 1991; 8(1-2): 171-7.
[http://dx.doi.org/10.1016/0267-6605(91)90027-D] [PMID: 10149164]

[22]   Chen WY, Abatangelo G. Functions of hyaluronan in wound repair. Wound Repair Regen 1999; 7(2): 79-89.
[http://dx.doi.org/10.1046/j.1524-475X.1999.00079.x] [PMID: 10231509]

[23]   Sneath RJ, Mangham DC. The normal structure and function of CD44 and its role in neoplasia. MP, Mol Pathol 1998; 51(4): 191-200.
[http://dx.doi.org/10.1136/mp.51.4.191] [PMID: 9893744]

[24]   Tokita Y, Okamoto A. Hydrolytic degradation of hyaluronic acid. Polym Degrad Stabil 1995; 48(2): 269-73.
[http://dx.doi.org/10.1016/0141-3910(95)00041-J]

[25]   Coradini D, Pellizzaro C, Miglierini G, Daidone MG, Perbellini A. Hyaluronic acid as drug delivery for sodium butyrate: improvement of the anti-proliferative activity on a breast-cancer cell line. Int J Cancer 1999; 81(3): 411-6.
[http://dx.doi.org/10.1002/(SICI)1097-0215(19990505)81:3<411::AID-IJC15>3.0.CO;2-F]   [PMID: 10209956]

[26]   Mekhail TM, Markman M. Paclitaxel in cancer therapy. Expert Opin Pharmacother 2002; 3(6): 755-66.
[http://dx.doi.org/10.1517/14656566.3.6.755] [PMID: 12036415]

[27]   Luo Y, Ziebell MR, Prestwich GD. A hyaluronic acid-taxol antitumor bioconjugate targeted to cancer cells. Biomacromolecules 2000; 1(2): 208-18.
[http://dx.doi.org/10.1021/bm000283n] [PMID: 11710102]

[28]   Thierry B, Kujawa P, Tkaczyk C, Winnik FM, Bilodeau L, Tabrizian M. Delivery platform for hydrophobic drugs: prodrug approach combined with self-assembled multilayers. J Am Chem Soc 2005; 127(6): 1626-7.
[http://dx.doi.org/10.1021/ja045077s] [PMID: 15700982]

[29]   Leonelli F, La Bella A, Francescangeli A, *et al.* A new and simply available class of hydrosoluble bioconjugates by coupling paclitaxel to hyaluronic acid through a 4-hydroxybutanoic acid derived linker. Helv Chim Acta 2005; 88(1): 154-9.
[http://dx.doi.org/10.1002/hlca.200490289]

[30]   Capodilupo A, Crescenzi V, Francescangeli A, *et al.* Hydrosoluble, metabolically fragile bioconjugates by coupling tetrabutylammonium hyaluronan with 2'-paclitaxel-4-bromobutyrate: Synthesis and antitumor properties. Hyaluronan 2003; 1: 391-5.

[31]   Montagner IM, Banzato A, Zuccolotto G, *et al.* Paclitaxel-hyaluronan hydrosoluble bioconjugate: mechanism of action in human bladder cancer cell lines. Urol Oncol 2013; 31(7): 1261-9.

[http://dx.doi.org/10.1016/j.urolonc.2012.01.005] [PMID: 22341413]

[32]   De Stefano I, Battaglia A, Zannoni GF, *et al.* Hyaluronic acid-paclitaxel: effects of intraperitoneal administration against CD44(+) human ovarian cancer xenografts. Cancer Chemother Pharmacol 2011; 68(1): 107-16.
[http://dx.doi.org/10.1007/s00280-010-1462-2] [PMID: 20848284]

[33]   Campisi M, Renier D, Pierimarchi P, Serafino A. Therapeutic use of new pharmaceutical preparations containing antitumoral drugs bound to hyaluronic acid in the treatment of neoplasias, European Patent 2009; 2279006.

[34]   Yin S, Huai J, Chen X, *et al.* Intracellular delivery and antitumor effects of a redox-responsive polymeric paclitaxel conjugate based on hyaluronic acid. Acta Biomater 2015; 26: 274-85.
[http://dx.doi.org/10.1016/j.actbio.2015.08.029] [PMID: 26300335]

[35]   Tringali G, Bettella F, Greco MC, Campisi M, Renier D, Navarra P. Pharmacokinetic profile of Oncofid-S after intraperitoneal and intravenous administration in the rat. J Pharm Pharmacol 2012; 64(3): 360-5.
[http://dx.doi.org/10.1111/j.2042-7158.2011.01417.x] [PMID: 22272720]

[36]   Serafino A, Zonfrillo M, Andreola F, *et al.* CD44-targeting for antitumor drug delivery: a new SN-3--hyaluronan bioconjugate for locoregional treatment of peritoneal carcinomatosis. Curr Cancer Drug Targets 2011; 11(5): 572-85.
[http://dx.doi.org/10.2174/156800911795655976] [PMID: 21486216]

[37]   Cai S, Thati S, Bagby TR, *et al.* Localized doxorubicin chemotherapy with a biopolymeric nanocarrier improves survival and reduces toxicity in xenografts of human breast cancer. J Control Release 2010; 146(2): 212-8.
[http://dx.doi.org/10.1016/j.jconrel.2010.04.006] [PMID: 20403395]

[38]   Dong Z, Zheng W, Xu Z, Yin Z. Improved stability and tumor targeting of 5-fluorouracil by conjugation with hyaluronan. J Appl Polym Sci 2013; 130(2): 927-32.
[http://dx.doi.org/10.1002/app.39247]

[39]   Sorbi C, Bergamin M, Bosi S, *et al.* Synthesis of 6-O-methotrexylhyaluronan as a drug delivery system. Carbohydr Res 2009; 344(1): 91-7.
[http://dx.doi.org/10.1016/j.carres.2008.09.021] [PMID: 18926524]

[40]   Homma A, Sato H, Okamachi A, *et al.* Novel hyaluronic acid-methotrexate conjugates for osteoarthritis treatment. Bioorg Med Chem 2009; 17(13): 4647-56.
[http://dx.doi.org/10.1016/j.bmc.2009.04.063] [PMID: 19457673]

[41]   Cai S, Xie Y, Bagby TR, Cohen MS, Forrest ML. Intralymphatic chemotherapy using a hyaluronan-cisplatin conjugate. J Surg Res 2008; 147(2): 247-52.
[http://dx.doi.org/10.1016/j.jss.2008.02.048] [PMID: 18498877]

[42]   Xie Y, Aillon KL, Cai S, *et al.* Pulmonary delivery of cisplatin-hyaluronan conjugates *via* endotracheal instillation for the treatment of lung cancer. Int J Pharm 2010; 392(1-2): 156-63.
[http://dx.doi.org/10.1016/j.ijpharm.2010.03.058] [PMID: 20363303]

[43]   Quan YH, Kim B, Park JH, Choi Y, Choi YH, Kim HK. Highly sensitive and selective anticancer effect by conjugated HA-cisplatin in non-small cell lung cancer overexpressed with CD44. Exp Lung

Res 2014; 40(10): 475-84.
[http://dx.doi.org/10.3109/01902148.2014.905656] [PMID: 25299431]

[44] Yang Q, Aires DJ, Cai S, *et al. In vivo* efficacy of nano hyaluronan-conjugated cisplatin for treatment of murine melanoma. J Drugs Dermatol 2014; 13(3): 283-7.
[PMID: 24595572]

[45] Ling X, Shen Y, Sun R, *et al.* Tumor-targeting delivery of hyaluronic acid–platinum (iv) nanoconjugate to reduce toxicity and improve survival. Polym Chem 2015; 6(9): 1541-52.
[http://dx.doi.org/10.1039/C4PY01592D]

[46] Della Valle F, Romeo A. Esters of hyaluronic acid. US Patent 4851521, 1989.

[47] Pouyani T, Prestwich GD. Functionalized derivatives of hyaluronic acid oligosaccharides: drug carriers and novel biomaterials. Bioconjug Chem 1994; 5(4): 339-47.
[http://dx.doi.org/10.1021/bc00028a010] [PMID: 7948100]

[48] Ferguson EL, Alshame AM, Thomas DW. Evaluation of hyaluronic acid-protein conjugates for polymer masked-unmasked protein therapy. Int J Pharm 2010; 402(1-2): 95-102.
[http://dx.doi.org/10.1016/j.ijpharm.2010.09.029] [PMID: 20888405]

[49] Mero A, Pasqualin M, Campisi M, Renier D, Pasut G. Conjugation of hyaluronan to proteins. Carbohydr Polym 2013; 92(2): 2163-70.
[http://dx.doi.org/10.1016/j.carbpol.2012.11.090] [PMID: 23399272]

[50] Sondergaard BC, Madsen SH, Segovia-Silvestre T, *et al.* Investigation of the direct effects of salmon calcitonin on human osteoarthritic chondrocytes. BMC Musculoskelet Disord 2010; 11(62): 62-247--11-62.
[http://dx.doi.org/10.1186/1471-2474-11-62]

[51] Mero A, Campisi M, Favero M, *et al.* A hyaluronic acid-salmon calcitonin conjugate for the local treatment of osteoarthritis: chondro-protective effect in a rabbit model of early OA. J Control Release 2014; 187: 30-8.
[http://dx.doi.org/10.1016/j.jconrel.2014.05.008] [PMID: 24837189]

[52] Kong JH, Oh EJ, Chae SY, Lee KC, Hahn SK. Long acting hyaluronateexendin 4 conjugate for the treatment of type 2 diabetes. Biomaterials 2010; 31(14): 4121-8.
[http://dx.doi.org/10.1016/j.biomaterials.2010.01.091] [PMID: 20149450]

[53] Hammes HP, Lin J, Bretzel RG, Brownlee M, Breier G. Upregulation of the vascular endothelial growth factor/vascular endothelial growth factor receptor system in experimental background diabetic retinopathy of the rat. Diabetes 1998; 47(3): 401-6.
[http://dx.doi.org/10.2337/diabetes.47.3.401] [PMID: 9519746]

[54] Oh EJ, Choi JS, Kim H, Joo CK, Hahn SK. Anti-Flt1 peptide - hyaluronate conjugate for the treatment of retinal neovascularization and diabetic retinopathy. Biomaterials 2011; 32(11): 3115-23.
[http://dx.doi.org/10.1016/j.biomaterials.2011.01.003] [PMID: 21277020]

[55] Pasut G, Veronese FM. State of the art in PEGylation: the great versatility achieved after forty years of research. J Control Release 2012; 161(2): 461-72.
[http://dx.doi.org/10.1016/j.jconrel.2011.10.037] [PMID: 22094104]

[56] Yang JA, Park K, Jung H, *et al.* Target specific hyaluronic acid-interferon alpha conjugate for the

treatment of hepatitis C virus infection. Biomaterials 2011; 32(33): 8722-9.
[http://dx.doi.org/10.1016/j.biomaterials.2011.07.088] [PMID: 21872329]

[57]   Yang JA, Kim ES, Kwon JH, *et al.* Transdermal delivery of hyaluronic acid human growth hormone conjugate. Biomaterials 2012; 33(25): 5947-54.
[http://dx.doi.org/10.1016/j.biomaterials.2012.05.003] [PMID: 22632765]

[58]   Friedrich EE, Sun LT, Natesan S, Zamora DO, Christy RJ, Washburn NR. Effects of hyaluronic acid conjugation on anti-TNF-α inhibition of inflammation in burns. J Biomed Mater Res A 2014; 102(5): 1527-36.
[http://dx.doi.org/10.1002/jbm.a.34829] [PMID: 23765644]

[59]   Mero A, Campisi M, Caputo M, *et al.* Hyaluronic acid as a protein polymeric carrier: An overview and a report on human growth hormone. Curr Drug Targets 2015; 16(13): 1503-11.
[http://dx.doi.org/10.2174/1389450116666150107151906] [PMID: 25563593]

[60]   Lee YH, Yoon HY, Shin JM, *et al.* A polymeric conjugate foreignizing tumor cells for targeted immunotherapy *in vivo.* J Control Release 2015; 199: 98-105.
[http://dx.doi.org/10.1016/j.jconrel.2014.12.007] [PMID: 25499555]

[61]   Varghese OP, Sun W, Hilborn J, Ossipov DA. *In situ* cross-linkable high molecular weight hyaluronan-bisphosphonate conjugate for localized delivery and cell-specific targeting: a hydrogel linked prodrug approach. J Am Chem Soc 2009; 131(25): 8781-3.
[http://dx.doi.org/10.1021/ja902857b] [PMID: 19499915]

[62]   Coradini D, Pellizzaro C, Scarlata I, *et al.* A novel retinoic/butyric hyaluronan ester for the treatment of acute promyelocytic leukemia: preliminary preclinical results. Leukemia 2006; 20(5): 785-92.
[http://dx.doi.org/10.1038/sj.leu.2404179] [PMID: 16525489]

[63]   Speranza A, Pellizzaro C, Coradini D. Hyaluronic acid butyric esters in cancer therapy. Anticancer Drugs 2005; 16(4): 373-9.
[http://dx.doi.org/10.1097/00001813-200504000-00003] [PMID: 15746573]

[64]   Coradini D, Pellizzaro C, Abolafio G, *et al.* Hyaluronic-acid butyric esters as promising antineoplastic agents in human lung carcinoma: a preclinical study. Invest New Drugs 2004; 22(3): 207-17.
[http://dx.doi.org/10.1023/B:DRUG.0000026247.72656.8a] [PMID: 15122068]

[65]   Rosato A, Banzato A, De Luca G, *et al.* HYTAD1-p20: a new paclitaxel-hyaluronic acid hydrosoluble bioconjugate for treatment of superficial bladder cancer. Urol Oncol 2006; 24(3): 207-15.
[http://dx.doi.org/10.1016/j.urolonc.2005.08.020] [PMID: 16678050]

[66]   Banzato A, Bobisse S, Rondina M, *et al.* A paclitaxel-hyaluronan bioconjugate targeting ovarian cancer affords a potent *in vivo* therapeutic activity. Clin Cancer Res 2008; 14(11): 3598-606.
[http://dx.doi.org/10.1158/1078-0432.CCR-07-2019] [PMID: 18519794]

[67]   Auzenne E, Ghosh SC, Khodadadian M, *et al.* Hyaluronic acid-paclitaxel: antitumor efficacy against CD44(+) human ovarian carcinoma xenografts. Neoplasia 2007; 9(6): 479-86.
[http://dx.doi.org/10.1593/neo.07229] [PMID: 17603630]

[68]   Lee SJ, Ghosh SC, Han HD, *et al.* Metronomic activity of CD44-targeted hyaluronic acid-paclitaxel in ovarian carcinoma. Clin Cancer Res 2012; 18(15): 4114-21.
[http://dx.doi.org/10.1158/1078-0432.CCR-11-3250] [PMID: 22693353]

[69]   Galer CE, Sano D, Ghosh SC, *et al.* Hyaluronic acid-paclitaxel conjugate inhibits growth of human squamous cell carcinomas of the head and neck *via* a hyaluronic acid-mediated mechanism. Oral Oncol 2011; 47(11): 1039-47.
[http://dx.doi.org/10.1016/j.oraloncology.2011.07.029] [PMID: 21903450]

[70]   Mittapalli RK, Liu X, Adkins CE, *et al.* Paclitaxel-hyaluronic nanoconjugates prolong overall survival in a preclinical brain metastases of breast cancer model. Mol Cancer Ther 2013; 12(11): 2389-99.
[http://dx.doi.org/10.1158/1535-7163.MCT-13-0132] [PMID: 24002934]

[71]   Montagner IM, Merlo A, Zuccolotto G, *et al.* Peritoneal tumor carcinomatosis: pharmacological targeting with hyaluronan-based bioconjugates overcomes therapeutic indications of current drugs. PLoS One 2014; 9(11): e112240.
[http://dx.doi.org/10.1371/journal.pone.0112240] [PMID: 25383653]

[72]   Oommen OP, Garousi J, Sloff M, Varghese OP. Tailored doxorubicin-hyaluronan conjugate as a potent anticancer glyco-drug: an alternative to prodrug approach. Macromol Biosci 2014; 14(3): 327-33.
[http://dx.doi.org/10.1002/mabi.201300383] [PMID: 24130147]

[73]   Cohen SM, Rockefeller N, Mukerji R, *et al.* Efficacy and toxicity of peritumoral delivery of nanoconjugated cisplatin in an *in vivo* murine model of head and neck squamous cell carcinoma. JAMA Otolaryngol Head Neck Surg 2013; 139(4): 382-7.
[http://dx.doi.org/10.1001/jamaoto.2013.214] [PMID: 23599074]

[74]   Venable RO, Worley DR, Gustafson DL, *et al.* Effects of intratumoral administration of a hyaluronan-cisplatin nanoconjugate to five dogs with soft tissue sarcomas. Am J Vet Res 2012; 73(12): 1969-76.
[http://dx.doi.org/10.2460/ajvr.73.12.1969] [PMID: 23176425]

[75]   Cohen SM, Mukerji R, Cai S, Damjanov I, Forrest ML, Cohen MS. Subcutaneous delivery of nanoconjugated doxorubicin and cisplatin for locally advanced breast cancer demonstrates improved efficacy and decreased toxicity at lower doses than standard systemic combination therapy *in vivo.* Am J Surg 2011; 202(6): 646-52.
[http://dx.doi.org/10.1016/j.amjsurg.2011.06.027] [PMID: 21982998]

[76]   Ling X, Shen Y, Sun R, *et al.* Tumor-targeting delivery of hyaluronic acid–platinum (iv) nanoconjugate to reduce toxicity and improve survival. Polym Chem 2015; 6(9): 1541-52.
[http://dx.doi.org/10.1039/C4PY01592D]

CHAPTER 5

Utilization of Glycosylation Engineering in the Development of Therapeutic Proteins

Robert Yite Chou[1], Kevin B. Turner[1] and Huijuan Li[2,*]

[1] *Sterile Product and Analytical Development, Biologics and Vaccine Development, Merck & Co., Inc., Kenilworth, NJ USA*

[2] *Moderna Therapeutics 200 Technology Square Cambridge, MA 02139, USA*

Abstract: Biologics are drugs made from complex molecules manufactured using living microorganisms, plant or animal cells. Many biologics are produced using recombinant DNA technology. Glycoproteins represent a major portion of biologics, including monoclonal antibodies, Fc-fusion proteins and other therapeutic proteins or enzymes. A thorough understanding of the nature and function of the carbohydrate moiety and its impact on pharmacology properties is essential in discovering, developing and manufacturing safe and efficacious glycoprotein biopharmaceuticals. This review summarizes the recent development in glycosylation engineering and characterization methodology. Examples of *N*-linked and/or *O*-linked glycosylation impacting drug pharmacology properties (including activity, pharmacokinetics, clearance, and immunogenicity) of marketed and developing therapeutic proteins are presented.

Keywords: ADCC, CHO, Fc-fusion, Fluorescence, Fucose, Glycan release, Glycoengineering, Glycoprotein, Glycosylation, Half-life, HILIC, HPLC, IgG, mAbs, Mass spectrometry, PAD, Pharmacodynamics, Pharmacokinetics, Recombinant human erythropoietin, Sialic acids.

INTRODUCTION

Achieving optimum therapeutic efficacy is dependent on maintaining a proper

* **Corresponding author Huijuan Li:** Moderna Therapeutics 200 Technology Square Cambridge, MA 02139, USA; Tel: 617-209-5850; Fax: 617-583-1998; E-mail: huijuan.li@modernatx.com

Qun Zhou (Ed.)

balance between drug exposure and its pharmacological effects. Therefore, the PK and PD parameters for therapeutics are often tuned through the drug design in a manner assuring that desired *in vivo* responses are achieved. PK refers to "what body does to the drug" and is influenced by drug absorption, distribution, metabolism and excretion, represented primarily by clearance rates, circulatory half-lives, volumes of distribution, and bioavailability. PD examines "what drug does to the body" and is influenced by the drug exposure and the drug pharmacological activities upon binding to the targets.

As a result of the intrinsic susceptibility of proteins to clearance mechanisms, protein drugs generally display limited plasma persistence. This can be overcome with higher protein concentrations and increased dosing frequencies. However, frequent treatment schedules coupled with the high target specificities and potencies can lead to unsuitably sharp dose/response profiles. The overstimulation of a targeted pathway can trigger autoregulatory feedback inhibition mechanisms that can lead to loss of *in vivo* efficacy. To overcome these limitations it has become routine practice to engineer the physicochemical and pharmacological properties of protein drugs in the early development.

Established technologies, such as targeted mutations, generation of fusion proteins and conjugates, glycosylation engineering, and pegylation, have been shown to significantly improve the efficacy of protein drugs by increasing their molecular stabilities and plasma persistence and consequently improving overall *in vivo* PD responses. Of these, engineered glycosylation is one of the most promising due to the fact that it has been shown to simultaneously improve the parameters necessary for therapeutic efficacy while maintaining target specificity.

The Processes of *N*- and *O*-linked Glycosylation

N-linked glycosylation is a post-translational modification that is conserved across eukaryotes. *N*-glycosylation begins in the endoplasmic reticulum (ER) with the transfer of a preassembled dolichol phosphate-linked fourteen sugar oligosaccharide to targeted asparagine residues in a protein [1, 2]. The preferred site for *N*-glycosylation contains the three amino acid sequence, Asn-Xaa-Ser/Thr where the second position can be any amino acid except Pro. This initial glycan transfer reaction is catalyzed by the heteromeric oligosaccharyltransferase (OST) complex and is supposed to precede folding of the protein in the ER [3]. Immediately after the oligosaccharide transfer, the two terminal glucose residues are cleaved off by α-glucosidase I and II and the resulting polypeptide with mono-glucosylated glycan structures ($Glc_1Man_9GlcNAc_2$) can interact with the ER-resident membrane-bound lectin calnexin or its soluble homolog calreticulin. These lectins support protein folding in a glycan-dependent protein quality control cycle. Secretory glycoproteins that have acquired their native conformation are released from the calnexin/calreticulin cycle and exit the ER to the Golgi apparatus. In the Golgi, the ER-derived oligomannosidic *N*-glycans on maturely folded glycoproteins are subjected to further *N*-glycan processing which generates the highly diverse complex N-glycans with different functional properties.

In contrast, *O*-linked glycosylation involves the covalent attachment of a monosaccharide to a Ser or Thr residue. Sequence specificity for guiding *O*-glycosylation has not been determined although it is presumed that folding and solubility of the protein play important roles in the position and number of *O*-linked glycans that occur on a given protein. Mammalian *O*-glycosylation is complex due to a variety of enzymes involved in multiple pathways resulting in a broad range of possible *O*-linked glycoforms. The most common form of human *O*-glycosylation occurs in the Golgi compartment and involves the transfer of a GalNAc residue from UDP-GalNAc to a Ser or Thr through the action of a GalNAc transferase followed by the subsequent transfer of any number of different sugars in different linkages including galactose, GalNAc and GlcNAc. Branching of the *O*-GalNAc chain, as well as the addition of lactosamine, also occurs providing additional oligosaccharide diversity [4]. The *O*-linked glycan chains can also be capped with sialic acid [5]. Different types of *O*-glycosylation (*e.g.*, *O*-linked fucose, glucose, mannose, xylose, or GalNAc) have been described on secretory proteins in mammals [6]. A class of *O*-linked glycosylation in mammals involving *O*-linked mannosyl species has been observed in brain, peripheral-nerve, and muscle glycoproteins [7]. The most well-studied of these, α-dystroglycan, is a heavily glycosylated extracellular membrane glycoprotein that functions to regulate important interactions with extracellular matrix proteins [8]. *O*-mannosylation on α-dystroglycan plays a critical role in normal physiology as illustrated clinically with the dystroglycanopathies [9].

Glycosylation Engineering

Glycosylation, the most common form of post-translational modification, is a diverse, enzyme mediated process by which oligosaccharide side chains are covalently attached to either the side chain of asparagine (*N*-linked) or serine/threonine (*O*-linked), with former being the most prominent. The oligosaccharide moiety of proteins is often essential for recognition, signaling, and interaction events within and between cells and proteins, and can play an important role in folding and defining the conformation of the protein. More than two-thirds of marketed therapeutic proteins, including monoclonal antibodies, are glycoproteins. Thus, while the biological function of most therapeutic proteins is typically defined principally by its templated amino acid component, the carbohydrate component of a therapeutic glycoprotein is often key in determining its pharmacological properties including stability, solubility/bioavailability, *in vivo* activity, pharmacokinetics, and immunogenicity. Most of the currently approved glycoproteins are produced in mammalian cell lines, which yield a heterogenous mixture of different glycoforms that closely resemble but are not fully identical to those in humans [10 - 14].

Glycoengineering is being developed as a method to control the composition of carbohydrates with many alternative glycoprotein production systems, such as mammalian, insect, yeast and plant cells, being actively investigated. Glycoengineering of cells for IgG1 expression has focused mainly on the elimination of core-fucose from the *N*-glycan at Asn297 in the Fc region of the heavy chain. The absence of core-fucose increases the affinity for FcγRIII receptor binding, leading to improved antibody-dependent cellular cytotoxicity on natural killer cells [15, 16]. Glycoengineered therapeutic antibodies, Mogamulizumab (anti-CCR4) and Obinutuzumab (anti-CD20), produced in core fucose-deficient CHO cells and core fucose-reduced CHO cells with GnTIII overexpression respectively, have been approved for human use by regulatory agencies and many others are in clinical trials [17, 21]. A similar glycosylation-dependent mechanism has an impact on antibody-dependent cellular phagocytosis by macrophages [17] and influences the receptor-mediated effector function of virus-neutralizing antibodies

[18]. ZMapp™ is composed of three "humanized" monoclonal antibodies manufactured in glycoengineered *Nicotiana benthamiana* plants that lack core-fucose residues. The ZMapp™ antibody cocktail is utilized for the treatment of Ebola virus infections and reversions of the disease in animal models [19]. In mice, the fucose-free monoclonal antibody 13F6, which is one of the ZMapp components, displayed clearly enhanced potency against Ebola virus compared to 13F6 variants with core fucose [20]. These examples clearly demonstrate the impact of glycosylation and highlight the potential of glycoengineered mAbs for different applications in humans. As a consequence, the number of glycoengineered mAbs approved for different treatments is expected to increase in the near future [21]. In addtion to modulating effector functions, engineering of mAbs for enhanced anti-inflammatory properties is an emerging glycosylation engineering strategy [22, 23]. IgG glycoforms with high amounts of terminal sialic acid can display increased affinity for lectin-type receptors and may represent another emerging field for the development of next-generation antibodies [24, 25]. In addition to antibody therapeutics, recombinant glycoprotein drugs with enhanced half-life, *i.e.* darbepoetin alfa, demonstrated that carbohydrate, especially high antennary sialylated glycoforms resulted in a prolonged and increased biological response *in vivo* [for review, see 26, 27].

Glycosylation Analysis of glycoproteins

In the development of therapeutic proteins, glycosylation analysis is performed to characterize the glycan profiles in glycoproteins to ensure robustness of processes and consistency of products. Especially for the biosimilar protein therapies, regulatory agencies have recommended to use appropriate analytical methodologies to compare the glycosylation profile of the proposed product and the reference product [28]. Beside the carbohydrate content (neutral sugars, amino sugars, and sialic acids), the structure of the carbohydrate chains, such as the oligosaccharide pattern (antennary profile) and the glycosylation site(s) of the protein, is also needed to be analyzed to the extent possible (ICH). The human IgG antibody with only one conserved *N*-glycosylation site at Asn297 in the Fc could generate up to 500 glycoforms [29]. Therefore, glycosylation analysis of glycoproteins can be very challenging.

Table 1. Methods used for characterization of glycans from glycoproteins.

Oligosaccharides	Motif	Analyte	Sample Prep to Separate Glycan from Protein	Sample Prep for Detection	Separation and Detection	Glycosylation Site Profiling	Quantity
N-Linked	amide group on Asn	Intact Glycoprotein	No	No	LC/MS	No	Semi
		Glycopeptides	Protease/ peptidase	No	LC/ MS	Yes	Semi
		Released Glycans/ Intact Glycans (complex, hybrid, oligomannose)	Glycosidase	Amine/Glycosylamine labeling (2-AA, 2-AB, iAB) for fluorescence detection [36]	HPLC-FD (HILIC)	No	Yes
				Permethylation to increase ionization efficiency and sensitivity [37]	LC/MS	Yes	Semi
				Stable isotopic amine (2-AA) labeling [36]	LC/MS	Yes	Yes
				Isolate (Cleaning Step)	HPAE-PAD	No	Yes
O-Linked	hydroxyl group on Ser/Thr	Released Glycans/ Intact Glycans	Alkaline β-Elimination/Hydrazinolysis[REMOVED REF FIELD] [38]/*O*-glycanase	No	LC/MS	Yes	Yes
Neutral Monosaccharide		Released Glycans/ Intact Glycans	Acid hydrolysis	Lyophilize	HPAE-PAD	No	Yes
Charged Monosaccharide/ Sialic Acids		Released Glycans/ Intact Glycans	Sialidase [40]	Isolate (Cleaning Step)	HPAE-PAD [39]	No	Yes
				Derivatization (2-AB [36], iAB) for fluorescence detection	HPAE-FD [36]	No	Yes

In general, there are three major strategies to perform the glycosylation analysis of glycoproteins, based on the type of analytes (summarized in Table **1**). The most common practice is to use either enzymatic or chemical cleavage to release glycans from polypeptide chains, and released glycans are further characterized, with or without prior derivatization and separation by high performance liquid chromatography (HPLC), followed by pulsed amperometric detection (PAD), fluorescence detection (FD), or mass spectrometry (MS) analysis [30]. As for profiling the glycosylation site(s), MS analyses of glycopeptides generated by specific proteases, the so-called bottom-up analysis, has become the preferred approach recently due to the advance of MS technology [31]. While there is no additional sample treatment for characterization of intact glycoprotein by MS, this top-down analysis was only applied to certain small glycoproteins with low levels of glycosylation [32]. For more complex glycoproteins with many isoforms, LC separation prior to MS analysis is required for intact glycoprotein profiling.

The release of glycans from glycoprotein by chemical procedures, such as hydrazinolysis or the β-elimination in an alkaline medium, has been recently replaced by the more gentle enzymatic deglycosylation, especially for *N*-glycans, due to the availability of PNGase F and PNGase A [33]. However, no specific enzyme toward *O*-glycans is available because of the diversity of *O*-linked core structures, except that unsubstituted core 1 strucutres may be removed by endo-α-N-acetylgalactosaminidase (*O*-glycanase) [33]. Therefore, alkaline β-elimination approach is still often used for the release of *O*-glycan, although *N*-glycans may also be cleaved in some situations and other post-translational modifications of protein may take place, such as deamidation of glutamine or asparagine [33].

Most glycans are complex structures with weak acidity or basicity, and lack chromophores or fluorophores, which make characterization challenging. The chemical derivatizations with fluorescent tags at the reducing end of the glycan not only make them separated easier by common HPLC separation methodologies through enhanced hydrophobicities, but also increase the detection sensitivities for optical detection (picomole range) and/or surge ionization efficiency for MS analysis (femtomole range) [34]. On the other hand, the protection of most or all functional groups for reducing polarity and for increasing thermal stability, such as permethylation, can improve MS ionization for increasing detection limits and providing linkage and branching information during fragmentation [35]. Although these hydrophobic tags making hydrophilic glycans appropriately suitable transit from the classical normal-phase liquid chromatography to the reverse-phase system, the lately advance in the amide polymeric column materials of the hydrophilic interaction chromatography (HILIC) is capable to well separate them based on the size and structure of glycans due to the more selective solute-solvent interactions [33], and enhances the efficiency of glycan analysis by further reducing the time for separation.

Glycosylation Related to Half-Life

It is well known that the removal of glycans from monoclonal antibodies (mAbs) can change the Fc domain conformation [40]. Studies have shown that CH2 domain of deglycosylated IgG exhibited the "closed" confirmation, as opposed to the "open" confirmation of the glycosylated version. Recent studies demonstrated that aglycosylated Fc possess a larger radius of gyration (Rg) than glycosylated Fc, suggesting that the proposed "closed" conformations of aglycosylated Fc observed in crystals may not be physiologically relevant [41]. The physical stability of deglycosylated IgG1 mAb by PNGaseF and recombinant, aglycosylated IgG1 Fc variants in comparison to the glycosylated versions have also been examined [42]. Relatively minor conformational changes of deglycosylated IgG1 mAbs or aglycosylated IgG1 Fc at lower temperature and near neutral pH were found, whereas significant conformational changes were observed in more extreme conditions such as at acidic pH and higher temperatures.

Conformational effects upon glycan removal do not appear to provide a meaningful impact on the PK properties of IgG *in vivo* [43]. Genetically engineered mouse–human chimeric IgG1 and IgG3 antibodies demonstrated that the PK profile of aglycosylated IgG1 mAb with Asn297 mutation was almost identical to that of the glycosylated form, although aglycosylated IgG3 molecules showed relatively faster clearance [44]. Similarly, the impacts of glycosylation with tunicamycin as a glycan depleting tool found that serum concentrations of the deglycosylated mouse IgG2b antibodies were very similar to those of the glycosylated forms [45].

Similar PK properties observed between glycosylated and nonglycosylated IgG suggest that the primary factor determining the PK of an IgG is not affected by glycan removal [43]. While aglycosylated IgG mAb showed similar FcRn-binding affinity to that of the glycosylated versions, mutational analysis of IgG molecules demonstrated that the FcRn-binding site overlaps with the protein A-binding site [43, 46]. Therefore it is not surprising that aglycosylated IgG can still bind to protein A and *E. coli* produced IgG exhibits similar PK to that of glycosylated versions produced from mammalian cells [43, 46]. However, IgG clearance may also be mediated by certain receptors, such as the asialoglycoprotein and mannose receptors, that are known to bind and clear specific glycan types. The asialoglycoprotein receptors bind terminal Gal residues [47], and the mannose receptors clear glycoproteins with terminal Man or GlcNAc sugars [48, 49]. Literature results indicate that high mannose glycans of therapeutic IgGs do increase serum clearance in humans [50], likely due to binding to mannose receptors.

The effects of glycosylation are not limited to mAbs [51]. Over the years biopharmaceuticals have substantially been engineered to optimize their efficacy and safety profiles [52]. Following its original introduction in 1989, second and third generation variants of recombinant human erythropoietin (rhEPO) appeared on the market in 2001 and 2007 [53]. Compared to the original product, representing the recombinant version of human EPO, serum half-life has substantially been improved by introducing two additional *N*-glycosylation sites (second generation) and by conjugation to polyethylene glycol (third generation). Aranesp®(darbepoetin), a second generation rhEPO, is engineered with two additional glycosylation sites resulting in five *N*- and one *O*-glycosylation sites and a total carbohydrate mass of 51% [54]. *N*-glycosylation sites are predominantly occupied with complex tetra-antennary glycans bearing terminal sialic acid residues that regulate the half-life of EPO in the blood stream. Third generation rhEPO (continuous erythropoiesis receptor activator – CERA) contains a single methoxy-polyethylene glycol PEG polymer of ~30 kDa integrated *via* amide bonds with the amino groups of either the N-terminus or lysine residues [55]. PEGylation is known to extend the drugs residence in the body and is particularly powerful for small proteins which, due to molecular size, need half-life extension [56, 57]. More recently, rhEPO was expressed in glycoengineered *Pichia pastoris* and showed that glycosylation fidelity is maintained in large fermentation volumes and that the protein can be purified to high homogeneity [58, 59]. In order to increase the half-life of rhEPO, the purified protein was coupled to polyethylene glycol (PEG) and then compared to the currently marketed erythropoiesis stimulating agent, Aranesp®(darbepoetin). In *in vitro* cell proliferation assays the PEGylated protein was slightly, and the non-PEGylated protein was significantly more active than comparator. Pharmacodynamics as well as pharmacokinetic activity of PEGylated rhEPO in animals was comparable to that of Aranesp®.

Like EPO, other low molecular weight proteins such as, recombinant human interferon β-1a (IFN-β-1a) [60] and alpha 1-antitrypsin (A!AT) [61] exhibit relatively short half-lives. Human A1AT exhibits three N-glycosylation sites at which biantennary, triantennary and traces of tetraantennary N-glycans can be found. Proper glycosylation is essential for correct protein folding, protein stability, biological activity, specific interaction with receptors and in particular for its serum half-life. Carbohydrates may play a role in protecting the glycoprotein from proteolytic cleavage by masking protease cleavage sites, thus positively affecting the circulation time of proteins. These properties can be modulated by the degree and composition of attached carbohydrates, such as sialic acid. As a terminal monosaccharide found on complex-type N-glycans, sialic acids are mainly responsible for the negative charge of glycans, which contributes to extension of glycoprotein half-life.

IFN-β is a 166 amino acid glycoprotein with a single N-linked carbohydrate chain on Asn80 residue. Commercial IFN-β-1a preparations are made up of a distribution of all core fucosylated glycoforms differing in sialylation and antennarity degree. The most abundant is the biantennary, disialylated core fucosylated IFN-β glycoform. For IFN-β, it is generally accepted that the carbohydrate moiety is mainly involved in promoting solubility and stability of the protein. Recent *in vitro* bioassay responses revealed a correlation mainly with the glycan antennarity. It is therefore suggested that all glycoforms have biological activity and play a role in modulating the overall IFN-β biological activity with higher antennarity glycoforms being able to better sustain IFN-β-1a bioactivity over time.

Linking an Fc fragment to a therapeutic protein prolongs the half-life of the associated protein that would otherwise be rapidly cleared. The PK behaviors of Fc-fusion proteins depend not only on Fc, but also on the fusion partner and associated glycosylations [62]. IgG1Fc is the most commonly used Fc for Fc-fusion proteins with partner molecules ranging from large proteins to small peptides, which may be heavily glycosylated or not glycosylated at all [62].

The clearance of Fc-fusion proteins is significantly higher than that of whole IgG molecules [43]. The most likely reasons involve the reduced affinity of Fc to FcRn are due to the fusion, the role of glycans in the fusion partner, and receptor-mediated clearance by partner molecule. The binding affinity at pH 6.0 to FcRn of several Fc-fusion products compared to those of intact IgG products demonstrated that the binding affinity of Fc-fusion proteins to FcRn was significantly reduced [63]. While the FcRn binding is clearly important, glycosylation patterns may play a more important role in determining the *in vivo* clearance. The investigation of humanized yeast-produced TNFαRII-Fc-fusion molecules demonstrated that it was the extent of sialylation on the partner molecule, not the FcRn-binding affinity, which determined the clearance [64]. The exposure was highly correlated to the quantity of the sialylation on the receptor molecule with higher sialic acid content resulting in higher exposure.

Other Fc-fusion molecules also rely on sialylation in reducing the *in vivo* clearance [43]. BR3–Fc-fusion has multiple *O*-linked glycosylation sites, and it has been shown that changes of the sialic acid content have a direct relationship with the PK properties [65]. Interleukin 23 receptor Fc-fusion protein (IL-23R-Fc) produced in human embryonic kidney (HEK) cells and CHO cells have differential PK properties correlating with significantly lower content of total sialic acids found in fusion-proteins produced by HEK cells [66]. Similarly, human cytotoxic T lymphocyte-associated protein 4 (CTLA4)-Ig is an IgG1 Fc-fusion with a relatively high total sialic acid content exhibited significantly slower clearance than those with lower sialic acid content [67] and the removal of sialic acid resulted in rapid circulation clearance. This report also showed that the type of sialic acids also matters [68]. Unlike the products produced in CHO cells, which has predominately terminal N-acetylneuraminic acid (Neu5Ac, NANA), the products produced in NS0 cells have N-glycolylneuraminic acid (Neu5Gc, NGNA). When tested *in vivo*, murine CTLA4-Ig produced in NS0 cells exhibited much faster clearance than that produced in CHO cells [67]. Clearance differences were also observed with lymphocyte function-associated molecule 3 (LFA3)TIP. When produced in murine NS0 myeloma cell lines a significant reduction in half-life of the molecule was observed due to the lack of NANA, whereas CHO-derived materials had high levels of NANA [68].

As briefly discussed above, differences in cell lines or expression systems impact glycosylation and PK properties. Typically, mammalian cell lines have been chosen for the production of most glycoproteins destined to be used as biopharmaceuticals because the resulting glycan profile is most similar to that of human proteins, at least compared to alternative cell platforms [69]. Non-human mammalian cells such as CHO have been used predominantly for the production of biopharmaceuticals including monoclonal antibodies (mAbs). Although the glycosylation profile of these products is 'human-like' the possibility of immunogenic epitopes such as Gal α1,3-Gal residues (α-Gal) and N-glycolylneuraminic acid (Neu5Gc) exists [69]. Human cell lines have now been designed for high productivity of recombinant proteins and ensuring authentic glycosylation patterns. The control of glycosylation on such proteins is important for the efficacy of recombinant biopharmaceuticals as well as the immunogenic properties [70, 71].

Mouse cells have an α1,3-galactosyltransferase enzyme, which is inactive in humans, that produces glycans containing α-Gal [70, 72]. Neu5Gc, a terminal neuraminic acid, is common in all non-primate mammalian cells and is formed by the hydroxylation of the common human form of sialic acid N-acetylneuraminic acid (Neu5Ac) through an enzyme, CMP-Neu5Ac hydroxylase which is not expressed in humans [73]. While few adverse immunogenic reactions have been attributed to CHO-generated proteins, the enzymes needed to form these non-human epitopes are present in the cells. Normal humans can have circulating antibodies against these two common non-human epitopes depending upon pre-exposure to xenotypic proteins [69].

Glycosylation Effects on PD

Therapeutic monoclonal antibodies recognizing cell surface expressed antigens, can engage Fcγ receptors (FcγR) on effector cells (monocytes, macrophages, natural killer (NK) cells, neutrophils, eosinophils and dendritic cells), or bind to complement 1q (C1q), and elicit immune effector functions such as antibody-dependent cellular phagocytosis (ADCP), antibody-dependent cell-mediated cytotoxicity (ADCC) and complement-dependent cytotoxicity (CDC) [43, 74 - 76]. Recombinant antibodies are often engineered to achieve desired receptor binding affinities for best therapeutic results. All antibodies contain a single *N*-linked glycosylation in their constant region. For IgGs, the glycosylation is associated with Asn297 and the invariant glycosylation site suggests its importance in antibody functions [77]. In fact, the removal of the Fc carbohydrates on this single glycosylation site dramatically reduced the Fc interactions with FcγRs [73]. *N*-glycosylation is one of the most important post-translational modifications and often results in a remarkable heterogeneity of protein glycoforms [40]. Depending on the recombinant expression system, this may be a complex type, hybrid type or high-mannose type structure. The use of mammalian expression systems generally results in complex type biantennary oligosaccharides in the Fc portions. These glycans may have a core fucose and a bisecting *N*-acetylglucosamine and can vary in terminal galactose and sialic acid content [29, 71, 78, 79].

Decreases in core fucosylation can increase IgG binding to FcγRIIIa and enhance ADCC activity [80, 81]. The significantly enhanced ADCC activity *in vivo* can be attributed to the enhanced Fcγ RIIIa-binding affinity by the afucosylated IgG Fc; without significantly higher affinity, the small amount of therapeutic mAb could be competed out by the large pool of endogenous IgG for binding to Fcγ receptors, resulting in no therapeutic effect [43, 82]. Similarly, high-mannose content in an IgG mAb results in higher binding affinity to FcγRIIIa coupled with higher ADCC activity [43, 83 - 86]. On the other hand, it has been shown that sialylation interfered with binding to FcγRIIIa, leading to reduced ADCC [87]. Unlike the other two glycosylations, terminal galactose does not seem to play a role in the ADCC activity, although small changes can be detected after degalactosylation [43, 88 - 91].

In addition to Fc glycosylations, some IgGs contain additional *N*-glycans in the variable regions of the fragment antigen binding (Fab) portion [40]. *N*-linked glycosylation of IgG Fab is contingent on the presence of consensus amino acid motifs making up the so-called "*N*-linked glycosylation sites" [88]. Although a glycosylation site is required, it is not sufficient for the addition of a glycan. Estimates of the percentage of Fab-glycosylated IgGs in healthy individuals range from ~15 to 25%, depending on the experimental approach used [92]. Compared with IgG Fc glycans, IgG Fab glycans contain high percentages of bisecting GlcNAc, galactose, and sialic acid and low percentages of core fucose [93]. Several studies showed that the presence of *N*-linked glycosylation sites in the variable domains can increase or decrease antigen-binding affinity. In one example, the anti–α(1→6)-dextran Ab 14.6b.1 has a 10-fold higher affinity compared with a single amino acid mutant that lacks the *N*-linked glycosylation site [92, 93]. Insertion of alternative *N*-linked glycosylation sites by mutagenesis around the pre-existing *N*-linked glycosylation site differentially affects the antigen binding, depending on the location of Fab glycans [93]. In another example, a 100-fold reduction in binding to tetanus toxoid, diphtheria toxoid, and dsDNA was observed for the polyreactive human mAb CBGA1 upon blocking *N*-linked glycosylation during production by tunicamycin [94]. The removal of sialic acid residues from Fab glycans can also decrease antigen-binding affinity [95]. As mentioned above, there is no Galα1,3Gal-linked glycans in human due to the lack of a specific enzyme for its biosynthesis, α1,3-galactosyltransferase, while antibody against this epitope is present in human serum. Therefore, the α-Gal epitope was engineered into glycol-vaccine for cancer immunotherapy [96, 97].

CONCLUDING REMARKS

The ability to understand and control the functions of glycosylation is the critical point to break through the development of therapeutic proteins, as they are involved in a wide range of biological and physiological processes. The recent development of glycosylation engineering and glycoprotein analysis has prompted the advance of therapeutic glycoproteins with the quality by design (QbD) paradigm to control and monitor the glycosylation that involves the desired clinical activities within the design space of properly controlled processes. Nevertheless, the profound glyco-heterogeneity of therapeutic protein involving complex biological mechanisms is still the major challenge for the utilization of glycosylation engineering in the development of therapeutic proteins.

CONFLICT OF INTEREST

The authors confirm that they have no conflict of interest to declare for this publication.

ACKNOWLEDGEMENTS

Declared none.

REFERENCES

[1]   Burda P, Aebi M. The dolichol pathway of N-linked glycosylation. Biochim Biophys Acta 1999; 1426(2): 239-57.
[http://dx.doi.org/10.1016/S0304-4165(98)00127-5] [PMID: 9878760]

[2]   Knauer R, Lehle L. The oligosaccharyltransferase complex from yeast. Biochim Biophys Acta 1999; 1426(2): 259-73.
[http://dx.doi.org/10.1016/S0304-4165(98)00128-7] [PMID: 9878773]

[3]   Aebi M. N-linked protein glycosylation in the ER. Biochim Biophys Acta 2013; 1833: 2430-7.1833;
[http://dx.doi.org/10.1016/j.bbamcr.2013.04.001] [PMID: 23583305]

[4]   Müller S, Hanisch FG. Recombinant MUC1 probe authentically reflects cell-specific *O*-glycosylation profiles of endogenous breast cancer mucin. High density and prevalent core 2-based glycosylation. J Biol Chem 2002; 277(29): 26103-12.
[http://dx.doi.org/10.1074/jbc.M202921200] [PMID: 12000758]

[5]   Sewell R, Bäckström M, Dalziel M, *et al.* The ST6GalNAc-I sialyltransferase localizes throughout the Golgi and is responsible for the synthesis of the tumor-associated sialyl-Tn *O*-glycan in human breast cancer. J Biol Chem 2006; 281(6): 3586-94.
[http://dx.doi.org/10.1074/jbc.M511826200] [PMID: 16319059]

[6]   Bennett EP, Mandel U, Clausen H, Gerken TA, Fritz TA, Tabak LA. Control of mucin-type O-glycosylation: a classification of the polypeptide GalNAc-transferase gene family. Glycobiology 2012; 22(6): 736-56.
[http://dx.doi.org/10.1093/glycob/cwr182] [PMID: 22183981]

[7]   Endo T. *O*-mannosyl glycans in mammals. Biochim Biophys Acta 1999; 1473(1): 237-46.
[http://dx.doi.org/10.1016/S0304-4165(99)00182-8] [PMID: 10580142]

[8]   Barresi R, Campbell KP. Dystroglycan: from biosynthesis to pathogenesis of human disease. J Cell Sci 2006; 119(Pt 2): 199-207.
[http://dx.doi.org/10.1242/jcs.02814] [PMID: 16410545]

[9]   Muntoni F, Brockington M, Blake DJ, Torelli S, Brown SC. Defective glycosylation in muscular dystrophy. Lancet 2002; 360(9343): 1419-21.
[http://dx.doi.org/10.1016/S0140-6736(02)11397-3] [PMID: 12424008]

[10]   Gerngross TU. Advances in the production of human therapeutic proteins in yeasts and filamentous fungi. Nat Biotechnol 2004; 22(11): 1409-14.
[http://dx.doi.org/10.1038/nbt1028] [PMID: 15529166]

[11]   Sethuraman N, Stadheim TA. Challenges in therapeutic glycoprotein production. Curr Opin Biotechnol 2006; 17(4): 341-6.
[http://dx.doi.org/10.1016/j.copbio.2006.06.010] [PMID: 16828275]

[12]   Hamilton SR, Gerngross TU. Glycosylation engineering in yeast: the advent of fully humanized yeast. Curr Opin Biotechnol 2007; 18(5): 387-92.
[http://dx.doi.org/10.1016/j.copbio.2007.09.001] [PMID: 17951046]

[13]   Li H, dAnjou M. Pharmacological significance of glycosylation in therapeutic proteins. Curr Opin Biotechnol 2009; 20(6): 678-84.

[http://dx.doi.org/10.1016/j.copbio.2009.10.009] [PMID: 19892545]

[14]   Beck A, Cochet O, Wurch T. GlycoFis technology to control the glycosylation of recombinant therapeutic proteins. Expert Opin Drug Discov 2010; 5(1): 95-111.
[http://dx.doi.org/10.1517/17460440903413504] [PMID: 22823974]

[15]   Engineered glycoforms of an antineuroblastoma IgG1 with optimized antibody-dependent cellular cytotoxic activity. Nat Biotechnol 1999; 17: 176-80.
[http://dx.doi.org/10.1038/6179] [PMID: 10052355]

[16]   Ferrara C, Grau S, Jäger C, *et al.* Unique carbohydrate-carbohydrate interactions are required for high affinity binding between FcgammaRIII and antibodies lacking core fucose. Proc Natl Acad Sci USA 2011; 108(31): 12669-74.
[http://dx.doi.org/10.1073/pnas.1108455108] [PMID: 21768335]

[17]   Golay J, Da Roit F, Bologna L, *et al.* Glycoengineered CD20 antibody obinutuzumab activates neutrophils and mediates phagocytosis through CD16B more efficiently than rituximab. Blood 2013; 122(20): 3482-91.
[http://dx.doi.org/10.1182/blood-2013-05-504043] [PMID: 24106207]

[18]   Forthal DN, Gach JS, Landucci G, *et al.* Fc-glycosylation influences Fcγ receptor binding and cell-mediated anti-HIV activity of monoclonal antibody 2G12. J Immunol 2010; 185(11): 6876-82.
[http://dx.doi.org/10.4049/jimmunol.1002600] [PMID: 21041724]

[19]   Qiu X, Wong G, Audet J, *et al.* Reversion of advanced Ebola virus disease in nonhuman primates with ZMapp. Nature 2014; 514(7520): 47-53.
[PMID: 25171469]

[20]   Zeitlin L, Pettitt J, Scully C, *et al.* Enhanced potency of a fucose-free monoclonal antibody being developed as an Ebola virus immunoprotectant. Proc Natl Acad Sci USA 2011; 108(51): 20690-4.
[http://dx.doi.org/10.1073/pnas.1108360108] [PMID: 22143789]

[21]   Beck A, Reichert JM. Marketing approval of mogamulizumab: a triumph for glyco-engineering. MAbs 2012; 4(4): 419-25.
[http://dx.doi.org/10.4161/mabs.20996] [PMID: 22699226]

[22]   Kaneko Y, Nimmerjahn F, Ravetch JV. Anti-inflammatory activity of immunoglobulin G resulting from Fc sialylation. Science 2006; 313(5787): 670-3.
[http://dx.doi.org/10.1126/science.1129594] [PMID: 16888140]

[23]   Raju TS, Lang SE. Diversity in structure and functions of antibody sialylation in the Fc. Curr Opin Biotechnol 2014; 30: 147-52.
[http://dx.doi.org/10.1016/j.copbio.2014.06.014] [PMID: 25032906]

[24]   Washburn N, Schwab I, Ortiz D, *et al.* Controlled tetra-Fc sialylation of IVIg results in a drug candidate with consistent enhanced anti-inflammatory activity. Proc Natl Acad Sci USA 2015; 112(11): E1297-306.
[http://dx.doi.org/10.1073/pnas.1422481112] [PMID: 25733881]

[25]   Dicker M, Strasser R. Using glyco-engineering to produce therapeutic proteins. Expert Opin Biol Ther 2015; 15(10): 1501-16.
[http://dx.doi.org/10.1517/14712598.2015.1069271] [PMID: 26175280]

[26]   Sinclair AM. Erythropoiesis stimulating agents: approaches to modulate activity. Biologics 2013; 7: 161-74.
        [PMID: 23847411]

[27]   Fauser BC, Mannaerts BM, Devroey P, Leader A, Boime I, Baird DT. Advances in recombinant DNA technology: corifollitropin alfa, a hybrid molecule with sustained follicle-stimulating activity and reduced injection frequency. Hum Reprod Update 2009; 15(3): 309-21.
        [http://dx.doi.org/10.1093/humupd/dmn065] [PMID: 19182099]

[28]   Guidance for industry, scientific considerations in demonstrating biosimilarity to a reference product US department of health and human services, food and drug administration, center for drug evaluation and research (CDER), and center for biologics evaluation and research (CBER), ICH Q6B 2015. http://www.fda.gov/downloads/DrugsGuidanceComplianceRegulatoryInformation/ Guidances/UCM291128.pdf

[29]   Jefferis R. Glycosylation as a strategy to improve antibody-based therapeutics. Nat Rev Drug Discov 2009; 8(3): 226-34.
        [http://dx.doi.org/10.1038/nrd2804] [PMID: 19247305]

[30]   Mariño K, Bones J, Kattla JJ, Rudd PM. A systematic approach to protein glycosylation analysis: a path through the maze. Nat Chem Biol 2010; 6(10): 713-23.
        [http://dx.doi.org/10.1038/nchembio.437] [PMID: 20852609]

[31]   Reusch D, Haberger M, Falck D, *et al.* Comparison of methods for the analysis of therapeutic immunoglobulin G Fc-glycosylation profiles-Part 2: Mass spectrometric methods. MAbs 2015; 7(4): 732-42.
        [http://dx.doi.org/10.1080/19420862.2015.1045173] [PMID: 25996192]

[32]   Harvey DJ. Structural determination of N-linked glycans by matrix-assisted laser desorption/ionization and electrospray ionization mass spectrometry. Proteomics 2005; 5(7): 1774-86.
        [http://dx.doi.org/10.1002/pmic.200401248] [PMID: 15832364]

[33]   Alley WR Jr, Mann BF, Novotny MV. High-sensitivity analytical approaches for the structural characterization of glycoproteins. Chem Rev 2013; 113(4): 2668-732.
        [http://dx.doi.org/10.1021/cr3003714] [PMID: 23531120]

[34]   Harvey DJ. Derivatization of carbohydrates for analysis by chromatography; electrophoresis and mass spectrometry. J Chromatogr B Analyt Technol Biomed Life Sci 2011; 879(17-18): 1196-225.
        [http://dx.doi.org/10.1016/j.jchromb.2010.11.010] [PMID: 21145794]

[35]   Ashline DJ, Lapadula AJ, Liu YH, *et al.* Carbohydrate structural isomers analyzed by sequential mass spectrometry. Anal Chem 2007; 79(10): 3830-42.
        [http://dx.doi.org/10.1021/ac062383a] [PMID: 17397137]

[36]   Ruhaak LR, Zauner G, Huhn C, Bruggink C, Deelder AM, Wuhrer M. Glycan labeling strategies and their use in identification and quantification. Anal Bioanal Chem 2010; 397(8): 3457-81.
        [http://dx.doi.org/10.1007/s00216-010-3532-z] [PMID: 20225063]

[37]   Pan S, Aebersold R. Quantitative proteomics by stable isotope labeling and mass spectrometry. In: Matthiesen R, Ed. Mass spectrometry data analysis in proteomics. 2nd ed. New Jersey: Humana Press 2007; pp. 290-1.

[38] Huang Y, Mechref Y, Novotny M. Glycoprotein cleavage protocol for oligosaccharide analysis. US 20040096948 A1, 2004.

[39] Fernandes DL. Biopharmaceutical sialylation Eur Biopharm Rev: 2006; 100-4. Spring

[40] Reusch D, Tejada ML. Fc glycans of therapeutic antibodies as critical quality attributes. Glycobiology 2015; 25(12): 1325-34.
[http://dx.doi.org/10.1093/glycob/cwv065] [PMID: 26263923]

[41] Borrok MJ, Jung ST, Kang TH, Monzingo AF, Georgiou G. Revisiting the role of glycosylation in the structure of human IgG Fc. ACS Chem Biol 2012; 7(9): 1596-602.
[http://dx.doi.org/10.1021/cb300130k] [PMID: 22747430]

[42] Alsenaidy MA, Okbazghi SZ, Kim JH, *et al.* Physical stability comparisons of IgG1-Fc variants: effects of N-glycosylation site occupancy and Asp/Gln residues at site Asn 297. J Pharm Sci 2014; 103(6): 1613-27.
[http://dx.doi.org/10.1002/jps.23975] [PMID: 24740840]

[43] Liu L. Antibody glycosylation and its impact on the pharmacokinetics and pharmacodynamics of monoclonal antibodies and Fc-fusion proteins. J Pharm Sci 2015; 104(6): 1866-84.
[http://dx.doi.org/10.1002/jps.24444] [PMID: 25872915]

[44] Tao MH, Morrison SL. Studies of aglycosylated chimeric mouse-human IgG. Role of carbohydrate in the structure and effector functions mediated by the human IgG constant region. J Immunol 1989; 143(8): 2595-601.
[PMID: 2507634]

[45] Nose M, Wigzell H. Biological significance of carbohydrate chains on monoclonal antibodies. Proc Natl Acad Sci USA 1983; 80(21): 6632-6.
[http://dx.doi.org/10.1073/pnas.80.21.6632] [PMID: 6579549]

[46] Kim JK, Tsen MF, Ghetie V, Ward ES. Catabolism of the murine IgG1 molecule: evidence that both CH2-CH3 domain interfaces are required for persistence of IgG1 in the circulation of mice. Scand J Immunol 1994; 40(4): 457-65.
[http://dx.doi.org/10.1111/j.1365-3083.1994.tb03488.x] [PMID: 7939418]

[47] Stockert RJ. The asialoglycoprotein receptor: relationships between structure, function, and expression. Physiol Rev 1995; 75(3): 591-609.
[PMID: 7624395]

[48] Stahl PD. The macrophage mannose receptor: current status. Am J Respir Cell Mol Biol 1990; 2(4): 317-8.
[http://dx.doi.org/10.1165/ajrcmb/2.4.317] [PMID: 2182080]

[49] Jones AJ, Papac DI, Chin EH, *et al.* Selective clearance of glycoforms of a complex glycoprotein pharmaceutical caused by terminal N-acetylglucosamine is similar in humans and cynomolgus monkeys. Glycobiology 2007; 17(5): 529-40.
[http://dx.doi.org/10.1093/glycob/cwm017] [PMID: 17331977]

[50] Goetze AM, Liu YD, Zhang Z, *et al.* High-mannose glycans on the Fc region of therapeutic IgG antibodies increase serum clearance in humans. Glycobiology 2011; 21(7): 949-59.
[http://dx.doi.org/10.1093/glycob/cwr027] [PMID: 21421994]

[51]   Walsh G, Jefferis R. Post-translational modifications in the context of therapeutic proteins. Nat Biotechnol 2006; 24(10): 1241-52.
[http://dx.doi.org/10.1038/nbt1252] [PMID: 17033665]

[52]   Carter PJ. Introduction to current and future protein therapeutics: a protein engineering perspective. Exp Cell Res 2011; 317(9): 1261-9.
[http://dx.doi.org/10.1016/j.yexcr.2011.02.013] [PMID: 21371474]

[53]   Sandra K, Vandenheede I, Sandra P. Modern chromatographic and mass spectrometric techniques for protein biopharmaceutical characterization. J Chromatogr A 2014; 1335: 81-103.
[http://dx.doi.org/10.1016/j.chroma.2013.11.057] [PMID: 24365115]

[54]   Macdougall IC, Gray SJ, Elston O, *et al.* Pharmacokinetics of novel erythropoiesis stimulating protein compared with epoetin alfa in dialysis patients. J Am Soc Nephrol 1999; 10(11): 2392-5.
[PMID: 10541299]

[55]   Macdougall IC. CERA (Continuous Erythropoietin Receptor Activator): a new erythropoiesis-stimulating agent for the treatment of anemia. Curr Hematol Rep 2005; 4(6): 436-40.
[PMID: 16232379]

[56]   Jevsevar S, Kunstelj M, Porekar VG. PEGylation of therapeutic proteins. Biotechnol J 2010; 5(1): 113-28.
[http://dx.doi.org/10.1002/biot.200900218] [PMID: 20069580]

[57]   Chen C, Constantinou A, Deonarain M. Modulating antibody pharmacokinetics using hydrophilic polymers. Expert Opin Drug Deliv 2011; 8(9): 1221-36.
[http://dx.doi.org/10.1517/17425247.2011.602399] [PMID: 21854300]

[58]   Liu L, Li H, Hamilton SR, *et al.* The impact of sialic acids on the pharmacokinetics of a PEGylated erythropoietin. J Pharm Sci 2012; 101(12): 4414-8.
[http://dx.doi.org/10.1002/jps.23320] [PMID: 22987365]

[59]   Nett JH, Gomathinayagam S, Hamilton SR, *et al.* Optimization of erythropoietin production with controlled glycosylation-PEGylated erythropoietin produced in glycoengineered *Pichia pastoris*. J Biotechnol 2012; 157(1): 198-206.
[http://dx.doi.org/10.1016/j.jbiotec.2011.11.002] [PMID: 22100268]

[60]   Mastrangeli R, Rossi M, Mascia M, *et al. In vitro* biological characterization of IFN-β-1a major glycoforms. Glycobiology 2015; (1): 21-9.

[61]   Lusch A, Kaup M, Marx U, Tauber R, Blanchard V, Berger M. Development and analysis of alpha 1-antitrypsin neoglycoproteins: the impact of additional N-glycosylation sites on serum half-life. Mol Pharm 2013; 10(7): 2616-29.
[http://dx.doi.org/10.1021/mp400043r] [PMID: 23668542]

[62]   Czajkowsky DM, Hu J, Shao Z, Pleass RJ. Fc-fusion proteins: new developments and future perspectives. EMBO Mol Med 2012; 4(10): 1015-28.
[http://dx.doi.org/10.1002/emmm.201201379] [PMID: 22837174]

[63]   Suzuki T, Ishii-Watabe A, Tada M, *et al.* Importance of neonatal FcR in regulating the serum half-life of therapeutic proteins containing the Fc domain of human IgG1: a comparative study of the affinity of monoclonal antibodies and Fc-fusion proteins to human neonatal FcR. J Immunol 2010; 184(4): 1968-

76.
[http://dx.doi.org/10.4049/jimmunol.0903296] [PMID: 20083659]

[64] Liu L, Gomathinayagam S, Hamuro L, *et al*. The impact of glycosylation on the pharmacokinetics of a TNFR2:Fc fusion protein expressed in Glycoengineered *Pichia Pastoris*. Pharm Res 2013; 30(3): 803-12.
[http://dx.doi.org/10.1007/s11095-012-0921-3] [PMID: 23135825]

[65] Stefanich EG, Ren S, Danilenko DM, *et al*. Evidence for an asialoglycoprotein receptor on nonparenchymal cells for O-linked glycoproteins. J Pharmacol Exp Ther 2008; 327(2): 308-15.
[http://dx.doi.org/10.1124/jpet.108.142232] [PMID: 18728239]

[66] Suen KF, Turner MS, Gao F, *et al*. Transient expression of an IL-23R extracellular domain Fc fusion protein in CHO *vs*. HEK cells results in improved plasma exposure. Protein Expr Purif 2010; 71(1): 96-102.
[http://dx.doi.org/10.1016/j.pep.2009.12.015] [PMID: 20045465]

[67] Flesher AR, Marzowski J, Wang WC, Raff HV. Fluorophore-labeled carbohydrate analysis of immunoglobulin fusion proteins: Correlation of oligosaccharide content with *in vivo* clearance profile. Biotechnol Bioeng 1995; 46(5): 399-407.
[http://dx.doi.org/10.1002/bit.260460502] [PMID: 18623330]

[68] Meier W, Gill A, Rogge M, *et al*. Immunomodulation by LFA3TIP, an LFA-3/IgG1 fusion protein: cell line dependent glycosylation effects on pharmacokinetics and pharmacodynamic markers. Ther Immunol 1995; 2(3): 159-71.
[PMID: 8885134]

[69] Butler M, Spearman M. The choice of mammalian cell host and possibilities for glycosylation engineering. Curr Opin Biotechnol 2014; 30: 107-12.
[http://dx.doi.org/10.1016/j.copbio.2014.06.010] [PMID: 25005678]

[70] Brooks SA. Appropriate glycosylation of recombinant proteins for human use: implications of choice of expression system. Mol Biotechnol 2004; 28(3): 241-55.
[http://dx.doi.org/10.1385/MB:28:3:241] [PMID: 15542924]

[71] Durocher Y, Butler M. Expression systems for therapeutic glycoprotein production. Curr Opin Biotechnol 2009; 20(6): 700-7.
[http://dx.doi.org/10.1016/j.copbio.2009.10.008] [PMID: 19889531]

[72] Macher BA, Galili U. The Galalpha1,3Galbeta1,4GlcNAc-R (α-Gal) epitope: a carbohydrate of unique evolution and clinical relevance. Biochim Biophys Acta 2008; 1780(2): 75-88. [BBA].
[http://dx.doi.org/10.1016/j.bbagen.2007.11.003] [PMID: 18047841]

[73] Ghaderi D, Zhang M, Hurtado-Ziola N, Varki A. Production platforms for biotherapeutic glycoproteins. Occurrence, impact, and challenges of non-human sialylation. Biotechnol Genet Eng Rev 2012; 28(1): 147-75.
[http://dx.doi.org/10.5661/bger-28-147] [PMID: 22616486]

[74] Krapp S, Mimura Y, Jefferis R, Huber R, Sondermann P. Structural analysis of human IgG-Fc glycoforms reveals a correlation between glycosylation and structural integrity. J Mol Biol 2003; 325(5): 979-89.
[http://dx.doi.org/10.1016/S0022-2836(02)01250-0] [PMID: 12527303]

[75] Arnold JN, Wormald MR, Sim RB, Rudd PM, Dwek RA. The impact of glycosylation on the biological function and structure of human immunoglobulins. Annu Rev Immunol 2007; 25: 21-50. [http://dx.doi.org/10.1146/annurev.immunol.25.022106.141702] [PMID: 17029568]

[76] Nimmerjahn F, Ravetch JV. Translating basic mechanisms of IgG effector activity into next generation cancer therapies. Cancer Immun 2012; 12: 13. [PMID: 22896758]

[77] Abès R, Teillaud JL. Impact of Glycosylation on Effector Functions of Therapeutic IgG. Pharmaceuticals 2010; 3(1): 146-57. [http://dx.doi.org/10.3390/ph3010146]

[78] Parekh RB, Dwek RA, Sutton BJ, *et al.* Association of rheumatoid arthritis and primary osteoarthritis with changes in the glycosylation pattern of total serum IgG. Nature 1985; 316(6027): 452-7. [http://dx.doi.org/10.1038/316452a0] [PMID: 3927174]

[79] Jefferis R. Glycosylation of recombinant antibody therapeutics. Biotechnol Prog 2005; 21(1): 11-6. [http://dx.doi.org/10.1021/bp040016j] [PMID: 15903235]

[80] Okazaki A, Shoji-Hosaka E, Nakamura K, *et al.* Fucose depletion from human IgG1 oligosaccharide enhances binding enthalpy and association rate between IgG1 and FcgammaRIIIa. J Mol Biol 2004; 336(5): 1239-49. [http://dx.doi.org/10.1016/j.jmb.2004.01.007] [PMID: 15037082]

[81] Jiang XR, Song A, Bergelson S, *et al.* Advances in the assessment and control of the effector functions of therapeutic antibodies. Nat Rev Drug Discov 2011; 10(2): 101-11. [http://dx.doi.org/10.1038/nrd3365] [PMID: 21283105]

[82] Iida S, Misaka H, Inoue M, *et al.* Nonfucosylated therapeutic IgG1 antibody can evade the inhibitory effect of serum immunoglobulin G on antibody-dependent cellular cytotoxicity through its high binding to FcgammaRIIIa. Clin Cancer Res 2006; 12(9): 2879-87. [http://dx.doi.org/10.1158/1078-0432.CCR-05-2619] [PMID: 16675584]

[83] Kanda Y, Yamada T, Mori K, *et al.* Comparison of biological activity among nonfucosylated therapeutic IgG1 antibodies with three different N-linked Fc oligosaccharides: the high-mannose, hybrid, and complex types. Glycobiology 2007; 17(1): 104-18. [http://dx.doi.org/10.1093/glycob/cwl057] [PMID: 17012310]

[84] Yu M, Brown D, Reed C, *et al.* Production, characterization, and pharmacokinetic properties of antibodies with N-linked mannose-5 glycans. MAbs 2012; 4(4): 475-87. [http://dx.doi.org/10.4161/mabs.20737] [PMID: 22699308]

[85] Kanda Y, Yamane-Ohnuki N, Sakai N, *et al.* Comparison of cell lines for stable production of fucose-negative antibodies with enhanced ADCC. Biotechnol Bioeng 2006; 94(4): 680-8. [http://dx.doi.org/10.1002/bit.20880] [PMID: 16609957]

[86] Zhou Q, Shankara S, Roy A, *et al.* Development of a simple and rapid method for producing non-fucosylated oligomannose containing antibodies with increased effector function. Biotechnol Bioeng 2008; 99(3): 652-65. [http://dx.doi.org/10.1002/bit.21598] [PMID: 17680659]

[87] Scallon BJ, Tam SH, McCarthy SG, Cai AN, Raju TS. Higher levels of sialylated Fc glycans in

immunoglobulin G molecules can adversely impact functionality. Mol Immunol 2007; 44(7): 1524-34.
[http://dx.doi.org/10.1016/j.molimm.2006.09.005] [PMID: 17045339]

[88]   Hodoniczky J, Zheng YZ, James DC. Control of recombinant monoclonal antibody effector functions by Fc N-glycan remodeling *in vitro*. Biotechnol Prog 2005; 21(6): 1644-52.
[http://dx.doi.org/10.1021/bp050228w] [PMID: 16321047]

[89]   Boyd PN, Lines AC, Patel AK. The effect of the removal of sialic acid, galactose and total carbohydrate on the functional activity of Campath-1H. Mol Immunol 1995; 32(17-18): 1311-8.
[http://dx.doi.org/10.1016/0161-5890(95)00118-2] [PMID: 8643100]

[90]   Peipp M, Dechant M, Valerius T. Effector mechanisms of therapeutic antibodies against ErbB receptors. Curr Opin Immunol 2008; 20(4): 436-43.
[http://dx.doi.org/10.1016/j.coi.2008.05.012] [PMID: 18585454]

[91]   Raju TS. Terminal sugars of Fc glycans influence antibody effector functions of IgGs. Curr Opin Immunol 2008; 20(4): 471-8.
[http://dx.doi.org/10.1016/j.coi.2008.06.007] [PMID: 18606225]

[92]   van de Bovenkamp FS, Hafkenscheid L, Rispens T, Rombouts Y. The Emerging Importance of IgG Fab Glycosylation in Immunity. J Immunol 2016; 196(4): 1435-41.
[http://dx.doi.org/10.4049/jimmunol.1502136] [PMID: 26851295]

[93]   Wallick SC, Kabat EA, Morrison SL. Glycosylation of a VH residue of a monoclonal antibody against alpha (1----6) dextran increases its affinity for antigen. J Exp Med 1988; 168(3): 1099-109.
[http://dx.doi.org/10.1084/jem.168.3.1099] [PMID: 2459288]

[94]   Leibiger H, Wüstner D, Stigler RD, Marx U. Variable domain-linked oligosaccharides of a human monoclonal IgG: structure and influence on antigen binding. Biochem J 1999; 338(Pt 2): 529-38.
[http://dx.doi.org/10.1042/bj3380529] [PMID: 10024532]

[95]   Khurana S, Raghunathan V, Salunke DM. The variable domain glycosylation in a monoclonal antibody specific to GnRH modulates antigen binding. Biochem Biophys Res Commun 1997; 234(2): 465-9.
[http://dx.doi.org/10.1006/bbrc.1997.5929] [PMID: 9177294]

[96]   Deguchi T, Tanemura M, Miyoshi E, *et al.* Increased immunogenicity of tumor-associated antigen, mucin 1, engineered to express alpha-gal epitopes: a novel approach to immunotherapy in pancreatic cancer. Cancer Res 2010; 70(13): 5259-69.
[http://dx.doi.org/10.1158/0008-5472.CAN-09-4313] [PMID: 20530670]

[97]   Tanemura M, Miyoshi E, Nagano H, *et al.* Role of α-gal epitope/anti-Gal antibody reaction in immunotherapy and its clinical application in pancreatic cancer. Cancer Sci 2013; 104(3): 282-90.
[http://dx.doi.org/10.1111/cas.12084] [PMID: 23240661]

Recent Advances in Biotechnology, 2016, *Vol. 3*, 152-171

CHAPTER 6

Antibody Engineering through Fc Glycans for Improved Therapeutic Index

Qun Zhou* and **Huawei Qiu**

Protein Engineering, Sanofi, 5 Mountain Road, Framingham, MA 01701, United States

Abstract: The N-glycans from Asn-297 site (or N297 glycans) of antibodies play important roles in antibody function. There is enhanced antibody-dependent cellular cytotoxicity (ADCC) with antibodies containing non-fucosylated glycans, while high anti-inflammatory activity was observed with the antibodies having highly sialylated biantennary structure. Many different approaches have been applied to engineer the N297 glycans with various structures to improve therapeutic efficacy by increasing antibody function, including pro-inflammatory effector function for cancer or antiviral therapy and anti-inflammatory activity for autoimmune diseases. Furthermore, the involvement of various N297 glycan forms in antibody structure-function relationship has also been recently elucidated. The crystal structures of FcγRIIIA complexed with non-fucosylated Fc demonstrated a more favorable carbohydrate-carbohydrate interaction, which is required for high affinity binding and enhanced ADCC activity. The strong anti-inflammatory activity of highly sialylated antibody is related to its interaction with dendritic cell specific intercellular adhesion molecule-3-grabbing nonintegrin (DC-SIGN). In addition, the N297 glycans were also remodeled for site-specific antibody conjugation. Thus, antibody glycoengineering provides valuable approaches for modulating antibody function, leading to increased therapeutic index.

Keywords: ADCC, Anti-inflammatory activity, Antibody-dependent cellular cytotoxicity, Autoimmune diseases, Bisecting GlcNAc containing glycans, Cancer, CDC, Complement-dependent cytotoxicity, Dendritic cell specific intercellular adhesion molecule-3-grabbing nonintegrin, FcγRIIIA, Glycoengineering, High mannose-type glycans, Highly galactosylated glycans,

---

* **Corresponding author Qun Zhou:** Protein Engineering, Sanofi, 5 Mountain Road, Framingham, MA 01701, United States; Tel/Fax: 508-270-2599; E-mail:qun.zhou@sanofi.com

Highly sialylated glycans, Homogeneous and site-specific ADC, N-glycans, N297 glycans, Non-fucosylated glycans, Site-specific antibody conjugation, Therapeutic antibodies.

INTRODUCTION

The antibody (or immunoglobulin) belongs to a family of major serum glycoproteins produced by humoral immune system. Its function includes recognition of specific antigens (pathogens) to facilitate their neutralization or killing by other effector cells or serum complements. Among antibodies, IgG is the main isotype. It contains two identical heavy (H) and two identical light (L) chains in a Y-shape, which are held together by a combination of non-covalent interactions and covalent interchain disulfide bonds. Each IgG has Fc and Fab regions, in which two Fab arms contain complementarity-determining regions critical for high affinity interaction with the antigens. The Fc region containing C_H2 and C_H3 is important in binding to either Fcγ receptors (FcγRs) on effector cells or complements for killing pathogens, as well as in the interaction with FcRn for long serum half-life.

The antibody IgG also contains the highly conserved Asn-297 (N297) site with attached N-glycans located in central cavity. In contrast to those found in most glycoproteins, the N297 glycans in C_H2 domains of recombinant therapeutic antibody are only partially processed with core fucosylated biantennary complex-type structures containing zero or one galactose at non-reducing terminus as major species (G0F or G1F). Other glycans, such as the processed sialylated biantennary complex-type, or less processed high mannose-type forms are also present, but at much lower levels. This unique feature is likely due to the low accessibility of N297 glycans inside the C_H2 domains toward galactosyltransferase or sialyltransferase in Golgi apparatus of the cells even when the IgG Fc molecule is in "open" conformation [1].

Although relatively simple compared to those found in other glycoproteins, the N297 glycans play a critical role in antibody structure and function. When the glycans are absent, there are significant structural changes in Fc region, which switches from "open" to "closed" conformation [1]. The binding of aglycosylated antibodies to FcγRs and complements was abolished with reduced antibody effector function, including antibody-dependent cellular cytotoxicity (ADCC) and complement-dependent cytotoxicity (CDC) [2]. Antibody glycosylation also plays a critical role in stability [2]. Reduction in thermostability of antibodies was observed through sequential deglycosylation.

In addition, there were changes in N297 glycans under certain physiological or pathological conditions [3 - 12]. The glycan profiles vary with age and pregnancy [13]. The glycosylation of antibody is often altered in multiple diseases, especially the autoimmune diseases (Table **1**).

Table 1. The autoimmune disease associated alteration in N-glycans in antibody IgG in serum except those noted in multiple sclerosis.

Disease	Fucosylation	Galactosylation	Sialylation	Bisecting GlcNAc
Rheumatoid arthritis		↓ [9, 14 - 17]		
Osteoarthritis		↓ [17]		
Systemic lupus erythematosus	↓ [18]	↓ [18]	↓ [18]	↑ [18]
Ulcerative colitis		↓ [2]		↑ [2]
Crohn's disease		↓ [2]	↓ [2]	
Kawasaki disease			↓ [19]	
Guillain-Barre syndrom		↓ [12]	↓ [12]	
Multiple sclerosis	↑* [3, 11]	↓* [3, 11]		↑* [3, 11]
Lambert-Eaton myasthenic syndrome and myasthenia gravis		↓ [20]		
Coeliac disease		↓ [21]		
Granulomatosis with polyangiitis		↓↓** [11, 22]	↓** [11, 22]	↓** [22]
Microscopic polyangiitis		↓ [22]		

*: IgG in cerebrospinal fluid.
**: Disease associated autoantibody, anti-proteinase 3.

There are lower galactosylation and sialylation in the patients with those diseases than the control. However, it is unknown whether these changes are related to the mechanism of the disease. With significant progress made during last two decades in developing therapeutic antibodies for treatment of multiple diseases, including cancer and autoimmune diseases, the role of N297 glycans in recombinant antibody function is further realized and antibody glycoengineering has been applied extensively in therapeutic development. Since there are many excellent reviews on antibody glycosylation [23 - 26], the current chapter briefly discusses the structural and functional relationship of N297 glycans with focus on recent progress in glycoengineering.

GLYCOENGINEERING FOR ENHANCED ANTIBODY EFFECTOR FUNCTION

Antibody effector functions include ADCC and CDC, which are dependent on its Fc region for binding to effector cells and molecules. When an antibody binds to the antigen on target cells, the affinity of the immune complex toward FcγRs, especially the FcγRIIIA (CD16) on NK cells is increased, resulting in the release of perforin and protease from the effector cells to kill the target cells. ADCC enhancement through glycoengineering was first reported ~17 years ago with a bisecting GlcNAc (β1,4-linked GlcNAc to inner mannose of tri-mannose core) containing monoclonal antibody from cells overexpressing GlcNAc transferase III [4, 27] (Table **2**). Since the bisecting GlcNAc formation blocks the addition of α1,6-linked fucose to the first GlcNAc of N297 glycans (core fucosylation) by fucosyltransferase VIII, it is likely that the ADCC enhancement associated with bisecting GlcNAc-containing antibody is due to its low fucose level. Indeed, about 50- to 100-fold enhanced antibody potency through ADCC was observed in non-fucosylated antibody through lectin enrichment or when expressed from fucose-deficient mutant cells [28, 29] (Table **2**).

The impact of non-fucosylation on ADCC was confirmed in the antibodies expressed from cells in which fucosyltransferase VIII gene was knocked out [41]. The high ADCC activity was also observed in antibodies with high mannose-type N297 glycans, which do not contain core fucose as the core fucosylation through fucosyltransferase VIII requires the presence of a GlcNAc residue β1,2-linked to mannose in the α1,3 arm of the tri-mannose core [36, 39, 40] (Table **2**). Although there is little change in CDC for antibodies with either bisecting GlcNAc or non-fucosylated complex-type glycans, the high mannose-type glycan-containing antibody showed lower CDC through C1q binding. The FcγRIIIA binding and ADCC activity of antibodies containing homogeneous glycans with different structures were reported recently [37, 38]. Interestingly, fully galactosylated or α2,6-linked sialylated non-fucosylated antibodies were superior to antibodies with other glycoforms in the interaction with FcγRIIIA (Table **2**). However, the sialylated fucosylated antibody showed lower FcγRIIIA binding and ADCC than non-sialylated fucosylated antibody [33].

Table 2. The effect of N297 glycans on antibody function.

Glycan Name	Glycan Structure**	ADCC and/or FcγRIIIA Binding	CDC*** and/or C1q Binding	Anti-inflammatory Activity
Fucosylated biantennary with zero galactose, or G0F*		+	+	+
Fucosylated biantennary with one galactose, or G1F*		+	+	+
Fucosylated biantennary with two galactose, or G2F		++ [30, 31]	++ [32]	++ [5, 6]
α2,6 linked disialylated biantennary with fucose, or A2F (2,6 linked)		± [33]	ND****	++++ [34, 35]
Non-fucosylated biantennary with zero galactose, or G0		+++ [28, 29]	+ [36]	ND
Non-fucosylated biantennary with one galactose, or G1		+++ [28, 29]	+ [36]	ND
Non-fucosylated biantennary with two galactose, or G2		++++ [37, 38]	ND	ND
α2,6 linked disialylated biantennary without fucose, or A2 (2,6 linked)		++++ [37, 38]	ND	ND
Non-fucosylated bisected biantennary with zero galactose, or Bisected G0		+++ [4, 37]	ND	ND

(Table 2) contd.....

Glycan Name	Glycan Structure**	ADCC and/or FcγRIIIA Binding	CDC*** and/or C1q Binding	Anti-inflammatory Activity
Non-fucosylated bisected biantennary with one galactose, or Bisected G1		+++ [4]	ND	ND
Oligomannose 9, or Man9		+++ [36, 39, 40]	± [36, 39, 40]	ND
Oligomannose 3, or Man3		+++ [37, 38]	ND	ND

*: G0F and G1F are major glycan species found in recombinant antibody IgG.

**: Symbols represent ■ as N-acetyl-glucosamine (GlcNAc); ● mannose; ○ galactose; ▶ fucose; ◆ sialic acid.

***: CDC represents classical pathway, but not alternative or lectin pathways.

****: ND is referred to as "not determined".

The ADCC enhancement in non-fucosylated antibodies is achieved through increased binding affinity to human FcγRIIIA expressed on NK cells. Once bound to non-fucosylated antibodies, the NK cells were stimulated and showed enhanced activation of proximal FcγRIIIA signaling and downstream pathways, as well as enhanced cytoskeletal rearrangement and degranulation with increased capability in killing target cells [42]. It was also reported recently that non-fucosylation increases phagocytic and cytotoxic activities associated with monocytes and macrophages through enhanced binding to FcγRIIIA expressed on these cells [43]. Increased binding of non-fucosylated antibody to the receptor on monocytes and macrophages was observed in the presence of nonspecific, endogenous IgGs, which mimics physiologic conditions. The superior *in vivo* anticancer efficacy of non-fucosylated antibodies was also demonstrated in tumor models using human FcγRIIIA transgenic mice or human PBMC-engrafted mice [44, 45].

The mechanism of increased affinity of non-fucosylated antibody for FcγRIIIA, but not other FcγRs, was further investigated. Interestingly, there was only a subtle change in the crystal structure within the C_H2 domain of non-fucosylated Fc compared to fucosylated Fc [46 - 48]. It was initially proposed that the presence of an additional bond between Tyr-296 of the non-fucosylated Fc and Lys-128 of FcγRIIIA was responsible for increased affinity [49]. However, Ferrara *et al.* found that increased binding of non-fucosylated antibody depended on the glycans at Asn-162 (N162 glycans) of the receptor, which were only present in human FcγRIIIA, but not other FcγRs [46]. The crystal structures of FcγRIIIA complexed with non-fucosylated Fc demonstrated a more favorable carbohydrate-carbohydrate interaction, which was required for high affinity [48]. The receptor N162 glycans interacted with the C_H2 domain in non-fucosylated Fc through a combination of direct water-mediated carbohydrate-carbohydrate (N162 glycans on the receptor and Fc N297 glycans) and carbohydrate-protein contacts (N162 glycans and Fc protein backbone). In FcγRIIIA complexed with fucosylated Fc, the core fucose in N297 glycans interfered with the interaction of the glycans with approaching N162 glycans from the receptor, resulting in lower binding affinity.

Due to the potential therapeutic value of ADCC enhancement in antibody therapy, many efforts have been made in generating non-fucosylated antibodies by various approaches. They include knock-out or knock-in of the genes coding for relevant glycosyltransferases or glycosidases, metabolic inhibition of these enzymes expressing in cells, *in vitro* glycosylation of antibodies, and different transgenic systems. The non-fucosylated antibody was generated from Chinese hamster ovary cells (CHO) overexpressing β1,4-GlcNAc transferase III gene [4, 27]. It was also produced from CHO cells in which the α1,6-fucosyltransferase VIII gene was knocked out using either homologous recombination or zinc finger nuclease technology. The gene coding for a bacterial enzyme was knocked into CHO cells to deplete fucose biosynthesized from the *de novo* pathway, resulting in non-fucosylated antibodies [41, 50 - 52]. A fucose analogue in inhibiting α1,6-fucosyltransferase and kifunensine in inhibiting α-mannosidase I were also used in producing antibodies with non-fucosylated complex-type glycans or high mannose-type glycans without re-engineering the cell lines [36, 39, 40, 53, 54]. The glycoengineering of yeast *Pichia pastoris*, insect cells and the aquatic plant *Lemna minor* was reported for producing non-fucosylated antibodies [55 - 57]. Non-fucosylated IgG and Fc were also generated through *in vitro* chemoenzymatic glycoengineering by using endoglycosidase-catalyzed transglycosylation of a recombinant antibody and aminooxy-GlcNAc conjugated

Fc which contained engineered aldehyde tags [58, 59].

In addition to the core fucosylation, the impact of galactosylation was also reported with higher CDC activity and FcγRIIIA binding [30 - 32], while sialylation of fucosylated antibody reduced its FcγRIIIA binding and ADCC [33] (Table **2**). The mechanism related to these changes is currently unknown.

Glycoengineered non-fucosylated antibodies with enhanced effector function have reached clinics with the recent approval of mogamulizumab (anti-CCR4 antibody) in Japan for treatment of CCR4-positive adult T-cell leukemia-lymphoma (ATL) and obinutuzumab (a second-generation anti-CD20 antibody, GA101) in the US and Europe for treatment of previously untreated chronic lymphocytic leukemia (CTL) [60, 61]. Mogamulizumab is a non-fucosylated antibody produced from CHO cells lacking fucosyltransferase VIII, while obinutuzumab is a non-fucosylated antibody generated by overexpressing GlcNAc transferase III in CHO cells [62, 63]. In addition to glycoengineering, obinutuzumab is a type II anti-CD20 antibody which binds antigens in a completely different orientation from type I antibodies, such as rituximab, due to modified elbow-hinge sequences in the variable region. It showed higher ADCC as result of glycoengineering, as well as enhanced direct programmed cell death and lower CDC [64]. Based on results from clinical trials, obinutuzumab combined with chemotherapy demonstrated a better response in treating chronic lymphocytic leukemia (CLL) than rituximab with the same chemotherapy [65].

GLYCOENGINEERING FOR ENHANCED ANTIBODY ANTI-INFLAMMATORY ACTIVITY

In addition to the involvement of pro-inflammatory ADCC enhancement, antibody glycans also play an important role in anti-inflammatory activity. Intravenous immunoglobulin (IVIG) has been used extensively in clinics for treating hematological and immunological diseases at high doses (1-2 g/kg). It was discovered that lectin-enriched IVIG with α2,6-linked sialylated glycans showed ~10-fold enhancement in anti-inflammatory activity in the K/BxN serum-induced arthritis model [35] (Table **2**). The same enhancement was also observed with IVIG and recombinant Fc sialylated *in vitro* using α2,6-sialyltransferase, but not sialylated using α2,3-linked sialyltransferase, aglycosylated or desialylated IVIG [34].

The mechanism of enhanced activity was further investigated. It was found that sialylation resulted in reduced FcγRIIIA binding by the antibody, which switched its specificity to a C-type lectin-like receptor, specific intracellular adhesion molecule-grabbing nonintegrin R1 (SIGN-R1), expressed on macrophages in the mouse splenic marginal zone, and a human homolog protein, dendritic-cell-specific intercellular adhesion molecule-3-grabbing nonintegrin (DC-SIGN) on dendritic cells and macrophages [66]. The anti-inflammatory activity of sialylated IVIG was abolished in SIGN-R1$^{-/-}$ mice, but re-established with transgenic expression of DC-SIGN. Anthony *et al.* showed that the *in vivo* interaction of sialylated IVIG with DC-SIGN transgenically expressed in SIGN-R1-/- mice results in IL-33 production and expansion of IL-4-producing basophils promoting increased expression of inhibitory Fc receptor FcγRIIB on effector macrophages [67]. The high expression of FcγRIIB on effector macrophages suppressed K/BxN serum-induced arthritis.

The anti-inflammatory enhancement by sialylated IVIG was also observed in other autoimmune disease mouse models including immune thrombocytopenic purpura and epidermolysis bullosa acquisita [68, 69]. However, due to the lack of competition between sialylated IVIG and other known glycan ligands for DC-SIGN [70], it was likely that the binding of sialylated IVIG to DC-SIGN on dendritic cells and macrophages was through antibody conformational change, leading to a strong protein-protein interaction, instead of carbohydrate-protein interaction. Indeed, the crystal structure of disialylated IgG Fc showed increased conformational flexibility of the C_H2 domain [71]. Compared to non-sialylated Fc with "open" conformation, the disialylated Fc adopted both "open" and "closed" conformations with the sialic acid visible only on the 6-arm of glycans.

Although different impacts of sialylation in IVIG on anti-inflammatory activity were observed, a recent report demonstrated that its disialylation with minimal impurity plays a critical role in activity enhancement [69]. Washburn *et al.* developed a robust and scalable process in generating sialylated IVIG containing ~90% disialylated biantennary glycan [69]. The highly sialylated IVIG showed consistent and enhanced anti-inflammatory activity up to 10-fold higher than IVIG in multiple mouse models of autoimmune diseases, including collagen antibody-induced arthritis (CAIA), K/BxN serum induced arthritis, immune thrombocytopenic purpura, and a prophylactic model of skin blistering disease. A chemoenzymatic approach was also used with mutants of endo-β-*N*-acetylglucosaminidase (EndoS) from *Streptococcus pyogenes* in preparing highly sialylated IVIG with >90% disialylated biantennary glycan [58].

Besides sialylated IVIG, the glycan-hydrolyzed IgG prepared with Endo S was found to suppress inflammatory arthritis induced by anti-type II collagen antibody in mice [72]. It was suggested that the dominant mechanism of suppression by glycan-hydrolyzed IgG was mediated through disturbances in the formation of immune complexes within the target tissue (joints). It was reported that the antibody galactosylation also played roles in FcγRIIB-mediated inhibition of autoimmune diseases [5, 6].

GLYCOENGINEERING FOR SITE-SPECIFIC ANTIBODY CONJUGATION

In addition to antibody effector function as a mechanism in killing tumor cells, antibody-drug conjugation, which combines the specific recognition of tumor antigen through antibodies with highly potent anticancer drugs, has been used as an effective therapy for oncology. Two antibody-drug conjugates (ADC), brentuximab vedotin (anti-CD30 ADC) and trastuzumab emtansine (Anti-Her2 ADC), have been approved by regulatory agencies for treating anaplastic large cell lymphoma/Hodgkin's lymphoma and advanced Her2-positive breast cancer, respectively.

Although highly effective, these ADCs were produced using conventional conjugation approaches. The former was produced by conjugating the drug-linker MMAE to four out of eight cysteine residues from four partially-reduced interchain hinge disulfides, while the latter was prepared by coupling the drug-linker maytansine to four out of 30-40 lysine residues in the antibody. Therefore, these approaches can potentially generate conjugates with significant heterogeneity containing one hundred to one million different species. Thus, many second generation methods, including those using antibody glycoengineering, are being explored for preparing homogeneous and site-specific ADC.

Since N297 glycans are located in the C_H2 domain, they provide a unique site for specifically conjugating anticancer drug-linker without interfering with antigen or FcRn binding. High concentrations of sodium periodate (10 to 12.5mM) was used in oxidizing N297 glycans (mainly G0F and G1F) of IgG to generate aldehydes for subsequent conjugation of drug-linker using hydrazone chemistry [73 - 75]. However, the exact number of drug-linkers conjugated at an individual monosaccharide, such as fucose, galactose, mannose or GlcNAc present in G0F and G1F, could not be controlled, nor the antibody aggregation and oxidation resulting from the treatment of periodate at high concentration. Since sialic acid is more sensitive to periodate oxidation than other monosaccharides, *in vitro* sialylation with sialyltransferase and galactosyltransferase was used to introduce a single sialic acid on each N-297 glycan (monosialylated fucosylated biantennary glycan, A1F) [76]. The introduced sialic acid was then oxidized with low concentration of periodate (1mM), followed by conjugation with aminooxy drug-linkers, generating a glyco ADC with ~1.6 drug-linker per antibody molecule. The *in vitro* and *in vivo* anticancer activities of glyco ADC were evaluated with three different antibodies (trastuzumab and two anti-fibroblast activation protein) conjugated with two different drug-linkers (aminooxy version of MMAE and PEG8-Dol10). They showed low protein aggregation, strong target-selective cytotoxicity and effective growth suppression of xenografted tumors in SCID mice.

In addition to native monosaccharide or sialic acid, sugar analogues or derivatives were also used for site-specific antibody-drug conjugation. It was reported that 6-thiofucose can be efficiently incorporated into the N297 glycans in antibodies expressed from CHO cells, which were then conjugated to drug-linkers using thiol-maleimide chemistry [77]. The purified anti-CD70 or anti-CD30 antibodies containing 1.2 to 1.4 thio-fucose per antibody were reduced and re-oxidized, followed by conjugation with MMAE drug-linker with an ADC of 1.3 drugs per antibody molecule. The ADC conjugated through thio-fucose was found stable in plasma and showed high cytotoxicity *in vitro*.

Furthermore, sugar nucleotide analogues were also investigated for *in vitro* glycosylation followed by a direct conjugation. It was reported previously that 2-acetonyl-2-deoxy-galactose or 2-*N*-acetyl azide GalNAc was transferred to the non-reducing termini of N297 glycans with mutant glycosyltransferase including galactosyltransferase with a single mutation (Y289L) [78, 79]. These galactose or GalNAc derivatives were efficiently conjugated to bioactive agents, such as biotin and fluorescent probes, containing alkyne/cyclooctyne or aminooxy functional groups. It was recently demonstrated that anticancer drug-linkers can be conjugated to these galactose or GalNAc derivatives in antibodies, which showed strong *in vitro* cytotoxicity [80]. The same approach was also used in site-specifically radiolabeling antibodies for delivering diagnostic and therapeutic radioisotopes to tumors [81, 82].

Van Geel *et al.* reported a two-stage process, in which the N297 glycans were first trimmed and tagged with azide before the drug-linker was coupled using copper-free click chemistry [83]. In the first step, the antibody glycans were trimmed by an endoglycosidase, liberating the core GlcNAc which was used as the substrate for transferring a GalNAc azide derivative by mutant galactosyltransferase as described above. In the second step, GalNAc azide in the antibody was coupled to a range of bicyclononyne (BCN)-linker containing different drugs, including doxorubicin, MMAE, MMAF and maytansine, with ~2 drug linkers per antibody. The ADCs were found to be homogeneous and highly hydrolytically stable with negligible aggregation. They showed stronger potency *in vitro* and efficacy *in vivo* than an ADC prepared using conventional conjugation with the same antibody and payload components despite the lower DAR.

Besides the galactose or GalNAc analogues, an azide-containing sialic acid derivative was also introduced into an anti-CD22 antibody with sialyltransferase and conjugated with dibenzylcyclooctyonol (DIBO or DBCO) linked doxorubicin using click chemistry [82]. The site-specific ADC conjugated through sialic acid derivatives showed strong *in vitro* anticancer activity.

CONCLUSION

There has been significant progress in antibody glycoengineering during the last fifteen years. Although relatively simple compared to those found in other glycoproteins, the antibody N297 glycans have been engineered into various forms with enhanced antibody functions, including pro-inflammatory effector function for cancer and antiviral therapy, as well as anti-inflammatory activity against autoimmune diseases. The glycoengineered forms improved the interaction between the antibodies and Fc receptors, including FcγRIIIA or DC-SIGN and have minimum perturbation in the antibody structure. According to published results, the increased affinity of the glycoengineered antibody for FcγRIIIA and DC-SIGN is mediated through a favorable interaction between protein to protein or carbohydrate to carbohydrate, but not carbohydrate to protein/lectin. On the other hand, site-specific glycoengineering followed by conjugation provides alternative methods for generating ADCs with homogeneous product profiles. More work is required for comparing these methods with other conventional or site-specific approaches *in vivo* as well as in clinics to demonstrate their feasibility and benefit in treating cancer patients. In summary, glycoengineering provides a unique approach in modulating antibody function and in creating specific sites for selective conjugation of anticancer drugs without any change in amino acid sequence. It is likely that many more effective therapeutic antibodies will be produced through glycoengineering in the future.

CONFLICT OF INTEREST

The authors confirm that they have no conflict of interest to declare for this publication.

ACKNOWLEDGEMENTS

The authors would like to thank Michael Drzyzga for assistance in preparing this book chapter.

REFERENCES

[1]   Krapp S, Mimura Y, Jefferis R, Huber R, Sondermann P. Structural analysis of human IgG-Fc glycoforms reveals a correlation between glycosylation and structural integrity. J Mol Biol 2003; 325(5): 979-89.
[http://dx.doi.org/10.1016/S0022-2836(02)01250-0] [PMID: 12527303]

[2]   Trbojević Akmačić I, Ventham NT, Theodoratou E, *et al.* Inflammatory bowel disease associates with proinflammatory potential of the immunoglobulin G glycome. Inflamm Bowel Dis 2015; 21(6): 1237-

47.
[PMID: 25895110]

[3]   Wuhrer M, Selman MH, McDonnell LA, *et al.* Pro-inflammatory pattern of IgG1 Fc glycosylation in multiple sclerosis cerebrospinal fluid. J Neuroinflammation 2015; 12: 235.
[http://dx.doi.org/10.1186/s12974-015-0450-1] [PMID: 26683050]

[4]   Umaña P, Jean-Mairet J, Moudry R, Amstutz H, Bailey JE. Engineered glycoforms of an antineuroblastoma IgG1 with optimized antibody-dependent cellular cytotoxic activity. Nat Biotechnol 1999; 17(2): 176-80.
[http://dx.doi.org/10.1038/6179] [PMID: 10052355]

[5]   Karsten CM, Pandey MK, Figge J, *et al.* Anti-inflammatory activity of IgG1 mediated by Fc galactosylation and association of FcγRIIB and dectin-1. Nat Med 2012; 18(9): 1401-6.
[http://dx.doi.org/10.1038/nm.2862] [PMID: 22922409]

[6]   Yamada K, Ito K, Furukawa J, *et al.* Galactosylation of IgG1 modulates FcγRIIB-mediated inhibition of murine autoimmune hemolytic anemia. J Autoimmun 2013; 47: 104-10.
[http://dx.doi.org/10.1016/j.jaut.2013.09.001] [PMID: 24055197]

[7]   Ohmi Y, Ise W, Harazono A, *et al.* Sialylation converts arthritogenic IgG into inhibitors of collagen-induced arthritis. Nat Commun 2016; 7: 11205.
[http://dx.doi.org/10.1038/ncomms11205] [PMID: 27046227]

[8]   Sumar N, Isenberg DA, Bodman KB, *et al.* Reduction in IgG galactose in juvenile and adult onset rheumatoid arthritis measured by a lectin binding method and its relation to rheumatoid factor. Ann Rheum Dis 1991; 50(9): 607-10.
[http://dx.doi.org/10.1136/ard.50.9.607] [PMID: 1929582]

[9]   Axford JS, Sumar N, Alavi A, *et al.* Changes in normal glycosylation mechanisms in autoimmune rheumatic disease. J Clin Invest 1992; 89(3): 1021-31.
[http://dx.doi.org/10.1172/JCI115643] [PMID: 1347295]

[10]   Kumpel BM, Wang Y, Griffiths HL, Hadley AG, Rook GA. The biological activity of human monoclonal IgG anti-D is reduced by beta-galactosidase treatment. Hum Antibodies Hybridomas 1995; 6(3): 82-8.
[PMID: 8597627]

[11]   Wuhrer M, Stavenhagen K, Koeleman CA, *et al.* Skewed Fc glycosylation profiles of anti-proteinase 3 immunoglobulin G1 autoantibodies from granulomatosis with polyangiitis patients show low levels of bisection, galactosylation, and sialylation. J Proteome Res 2015; 14(4): 1657-65.
[http://dx.doi.org/10.1021/pr500780a] [PMID: 25761865]

[12]   Fokkink WJ, Selman MH, Dortland JR, *et al.* IgG Fc N-glycosylation in Guillain-Barré syndrome treated with immunoglobulins. J Proteome Res 2014; 13(3): 1722-30.
[http://dx.doi.org/10.1021/pr401213z] [PMID: 24533874]

[13]   Wuhrer M, Stam JC, van de Geijn FE, *et al.* Glycosylation profiling of immunoglobulin G (IgG) subclasses from human serum. Proteomics 2007; 7(22): 4070-81.
[http://dx.doi.org/10.1002/pmic.200700289] [PMID: 17994628]

[14]   Youings A, Chang SC, Dwek RA, Scragg IG. Site-specific glycosylation of human immunoglobulin G

is altered in four rheumatoid arthritis patients. Biochem J 1996; 314(Pt 2): 621-30.
[http://dx.doi.org/10.1042/bj3140621] [PMID: 8670078]

[15] Pekelharing JM, Hepp E, Kamerling JP, Gerwig GJ, Leijnse B. Alterations in carbohydrate composition of serum IgG from patients with rheumatoid arthritis and from pregnant women. Ann Rheum Dis 1988; 47(2): 91-5.
[http://dx.doi.org/10.1136/ard.47.2.91] [PMID: 3355256]

[16] Quast I, Keller CW, Maurer MA, *et al.* Sialylation of IgG Fc domain impairs complement-dependent cytotoxicity. J Clin Invest 2015; 125(11): 4160-70.
[http://dx.doi.org/10.1172/JCI82695] [PMID: 26436649]

[17] Parekh RB, Dwek RA, Sutton BJ, *et al.* Association of rheumatoid arthritis and primary osteoarthritis with changes in the glycosylation pattern of total serum IgG. Nature 1985; 316(6027): 452-7.
[http://dx.doi.org/10.1038/316452a0] [PMID: 3927174]

[18] Sjöwall C, Zapf J, von Löhneysen S, *et al.* Altered glycosylation of complexed native IgG molecules is associated with disease activity of systemic lupus erythematosus. Lupus 2015; 24(6): 569-81.
[http://dx.doi.org/10.1177/0961203314558861] [PMID: 25389233]

[19] Ogata S, Shimizu C, Franco A, *et al.* Treatment response in Kawasaki disease is associated with sialylation levels of endogenous but not therapeutic intravenous immunoglobulin G. PLoS One 2013; 8(12): e81448.
[http://dx.doi.org/10.1371/journal.pone.0081448] [PMID: 24324693]

[20] Selman MH, Niks EH, Titulaer MJ, Verschuuren JJ, Wuhrer M, Deelder AM. IgG fc N-glycosylation changes in Lambert-Eaton myasthenic syndrome and myasthenia gravis. J Proteome Res 2011; 10(1): 143-52.
[http://dx.doi.org/10.1021/pr1004373] [PMID: 20672848]

[21] Cremata JA, Sorell L, Montesino R, *et al.* Hypogalactosylation of serum IgG in patients with coeliac disease. Clin Exp Immunol 2003; 133(3): 422-9.
[http://dx.doi.org/10.1046/j.1365-2249.2003.02220.x] [PMID: 12930370]

[22] Holland M, Takada K, Okumoto T, *et al.* Hypogalactosylation of serum IgG in patients with ANCA-associated systemic vasculitis. Clin Exp Immunol 2002; 129(1): 183-90.
[http://dx.doi.org/10.1046/j.1365-2249.2002.01864.x] [PMID: 12100039]

[23] Jefferis R. Glycosylation as a strategy to improve antibody-based therapeutics. Nat Rev Drug Discov 2009; 8(3): 226-34.
[http://dx.doi.org/10.1038/nrd2804] [PMID: 19247305]

[24] Jefferis R. A sugar switch for anti-inflammatory antibodies. Nat Biotechnol 2006; 24(10): 1230-1.
[http://dx.doi.org/10.1038/nbt1006-1230] [PMID: 17033662]

[25] Anthony RM, Wermeling F, Ravetch JV. Novel roles for the IgG Fc glycan. Ann N Y Acad Sci 2012; 1253: 170-80.
[http://dx.doi.org/10.1111/j.1749-6632.2011.06305.x] [PMID: 22288459]

[26] Niwa R, Satoh M. The current status and prospects of antibody engineering for therapeutic use: focus on glycoengineering technology. J Pharm Sci 2015; 104(3): 930-41.
[http://dx.doi.org/10.1002/jps.24316] [PMID: 25583555]

[27]   Davies J, Jiang L, Pan LZ, LaBarre MJ, Anderson D, Reff M. Expression of GnTIII in a recombinant anti-CD20 CHO production cell line: Expression of antibodies with altered glycoforms leads to an increase in ADCC through higher affinity for FC gamma RIII. Biotechnol Bioeng 2001; 74(4): 288-94.
        [http://dx.doi.org/10.1002/bit.1119] [PMID: 11410853]

[28]   Shields RL, Lai J, Keck R, *et al.* Lack of fucose on human IgG1 N-linked oligosaccharide improves binding to human Fcgamma RIII and antibody-dependent cellular toxicity. J Biol Chem 2002; 277(30): 26733-40.
        [http://dx.doi.org/10.1074/jbc.M202069200] [PMID: 11986321]

[29]   Shinkawa T, Nakamura K, Yamane N, *et al.* The absence of fucose but not the presence of galactose or bisecting N-acetylglucosamine of human IgG1 complex-type oligosaccharides shows the critical role of enhancing antibody-dependent cellular cytotoxicity. J Biol Chem 2003; 278(5): 3466-73.
        [http://dx.doi.org/10.1074/jbc.M210665200] [PMID: 12427744]

[30]   Houde D, Peng Y, Berkowitz SA, Engen JR. Post-translational modifications differentially affect IgG1 conformation and receptor binding. Mol Cell Proteomics 2010; 9(8): 1716-28.
        [http://dx.doi.org/10.1074/mcp.M900540-MCP200] [PMID: 20103567]

[31]   Kumpel BM, Rademacher TW, Rook GA, Williams PJ, Wilson IB. Galactosylation of human IgG monoclonal anti-D produced by EBV-transformed B-lymphoblastoid cell lines is dependent on culture method and affects Fc receptor-mediated functional activity. Hum Antibodies Hybridomas 1994; 5(3-4): 143-51.
        [PMID: 7756579]

[32]   Raju TS. Terminal sugars of Fc glycans influence antibody effector functions of IgGs. Curr Opin Immunol 2008; 20(4): 471-8.
        [http://dx.doi.org/10.1016/j.coi.2008.06.007] [PMID: 18606225]

[33]   Scallon BJ, Tam SH, McCarthy SG, Cai AN, Raju TS. Higher levels of sialylated Fc glycans in immunoglobulin G molecules can adversely impact functionality. Mol Immunol 2007; 44(7): 1524-34.
        [http://dx.doi.org/10.1016/j.molimm.2006.09.005] [PMID: 17045339]

[34]   Anthony RM, Nimmerjahn F, Ashline DJ, Reinhold VN, Paulson JC, Ravetch JV. Recapitulation of IVIG anti-inflammatory activity with a recombinant IgG Fc. Science 2008; 320(5874): 373-6.
        [http://dx.doi.org/10.1126/science.1154315] [PMID: 18420934]

[35]   Kaneko Y, Nimmerjahn F, Ravetch JV. Anti-inflammatory activity of immunoglobulin G resulting from Fc sialylation. Science 2006; 313(5787): 670-3.
        [http://dx.doi.org/10.1126/science.1129594] [PMID: 16888140]

[36]   Kanda Y, Yamada T, Mori K, *et al.* Comparison of biological activity among nonfucosylated therapeutic IgG1 antibodies with three different N-linked Fc oligosaccharides: the high-mannose, hybrid, and complex types. Glycobiology 2007; 17(1): 104-18.
        [http://dx.doi.org/10.1093/glycob/cwl057] [PMID: 17012310]

[37]   Lin CW, Tsai MH, Li ST, *et al.* A common glycan structure on immunoglobulin G for enhancement of effector functions. Proc Natl Acad Sci USA 2015; 112(34): 10611-6.
        [http://dx.doi.org/10.1073/pnas.1513456112] [PMID: 26253764]

[38]   Kurogochi M, Mori M, Osumi K, *et al.* Glycoengineered monoclonal antibodies with homogeneous

glycan (M3, G0, G2, and A2) using a chemoenzymatic approach have different affinities for FcγRIIIa and variable antibody-dependent cellular cytotoxicity activities. PLoS One 2015; 10(7): e0132848.
[http://dx.doi.org/10.1371/journal.pone.0132848] [PMID: 26200113]

[39]   Zhou Q, Shankara S, Roy A, *et al.* Development of a simple and rapid method for producing non-fucosylated oligomannose containing antibodies with increased effector function. Biotechnol Bioeng 2008; 99(3): 652-65.
[http://dx.doi.org/10.1002/bit.21598] [PMID: 17680659]

[40]   Yu M, Brown D, Reed C, *et al.* Production, characterization, and pharmacokinetic properties of antibodies with N-linked mannose-5 glycans. MAbs 2012; 4(4): 475-87.
[http://dx.doi.org/10.4161/mabs.20737] [PMID: 22699308]

[41]   Yamane-Ohnuki N, Kinoshita S, Inoue-Urakubo M, *et al.* Establishment of FUT8 knockout Chinese hamster ovary cells: an ideal host cell line for producing completely defucosylated antibodies with enhanced antibody-dependent cellular cytotoxicity. Biotechnol Bioeng 2004; 87(5): 614-22.
[http://dx.doi.org/10.1002/bit.20151] [PMID: 15352059]

[42]   Liu SD, Chalouni C, Young JC, Junttila TT, Sliwkowski MX, Lowe JB. Afucosylated antibodies increase activation of FcγRIIIa-dependent signaling components to intensify processes promoting ADCC. Cancer Immunol Res 2015; 3(2): 173-83.
[http://dx.doi.org/10.1158/2326-6066.CIR-14-0125] [PMID: 25387893]

[43]   Herter S, Birk MC, Klein C, Gerdes C, Umana P, Bacac M. Glycoengineering of therapeutic antibodies enhances monocyte/macrophage-mediated phagocytosis and cytotoxicity. J Immunol 2014; 192(5): 2252-60.
[http://dx.doi.org/10.4049/jimmunol.1301249] [PMID: 24489098]

[44]   Junttila TT, Parsons K, Olsson C, *et al.* Superior *in vivo* efficacy of afucosylated trastuzumab in the treatment of HER2-amplified breast cancer. Cancer Res 2010; 70(11): 4481-9.
[http://dx.doi.org/10.1158/0008-5472.CAN-09-3704] [PMID: 20484044]

[45]   Niwa R, Shoji-Hosaka E, Sakurada M, *et al.* Defucosylated chimeric anti-CC chemokine receptor 4 IgG1 with enhanced antibody-dependent cellular cytotoxicity shows potent therapeutic activity to T-cell leukemia and lymphoma. Cancer Res 2004; 64(6): 2127-33.
[http://dx.doi.org/10.1158/0008-5472.CAN-03-2068] [PMID: 15026353]

[46]   Sondermann P, Brunker P, and Umana P., "The Carbohydrate at FcγRIIIa Asn-162 an element required for high affinity binding to non-fucosylated IgG glycoforms. J Biol Chem 2006; 281: 5032-6.
[http://dx.doi.org/10.1074/jbc.M510171200] [PMID: 16330541]

[47]   Matsumiya S, Yamaguchi Y, Saito J, *et al.* Structural comparison of fucosylated and nonfucosylated Fc fragments of human immunoglobulin G1. J Mol Biol 2007; 368(3): 767-79.
[http://dx.doi.org/10.1016/j.jmb.2007.02.034] [PMID: 17368483]

[48]   Ferrara C, Grau S, Jäger C, *et al.* Unique carbohydrate-carbohydrate interactions are required for high affinity binding between FcgammaRIII and antibodies lacking core fucose. Proc Natl Acad Sci USA 2011; 108(31): 12669-74.
[http://dx.doi.org/10.1073/pnas.1108455108] [PMID: 21768335]

[49]   Okazaki A, Shoji-Hosaka E, Nakamura K, *et al.* Fucose depletion from human IgG1 oligosaccharide enhances binding enthalpy and association rate between IgG1 and FcgammaRIIIa. J Mol Biol 2004;

336(5): 1239-49.
[http://dx.doi.org/10.1016/j.jmb.2004.01.007] [PMID: 15037082]

[50] Imai-Nishiya H, Mori K, Inoue M, *et al*. Double knockdown of alpha1,6-fucosyltransferase (FUT8) and GDP-mannose 4,6-dehydratase (GMD) in antibody-producing cells: a new strategy for generating fully non-fucosylated therapeutic antibodies with enhanced ADCC. BMC Biotechnol 2007; 7: 84.
[http://dx.doi.org/10.1186/1472-6750-7-84] [PMID: 18047682]

[51] Wong AW, Baginski TK, Reilly DE. Enhancement of DNA uptake in FUT8-deleted CHO cells for transient production of afucosylated antibodies. Biotechnol Bioeng 2010; 106(5): 751-63.
[http://dx.doi.org/10.1002/bit.22749] [PMID: 20564613]

[52] von Horsten HH, Ogorek C, Blanchard V, *et al*. Production of non-fucosylated antibodies by co-expression of heterologous GDP-6-deoxy-D-lyxo-4-hexulose reductase. Glycobiology 2010; 20(12): 1607-18.
[http://dx.doi.org/10.1093/glycob/cwq109] [PMID: 20639190]

[53] van Berkel PH, Gerritsen J, van Voskuilen E, *et al*. Rapid production of recombinant human IgG With improved ADCC effector function in a transient expression system. Biotechnol Bioeng 2010; 105(2): 350-7.
[http://dx.doi.org/10.1002/bit.22535] [PMID: 19739094]

[54] Okeley NM, Alley SC, Anderson ME, *et al*. Development of orally active inhibitors of protein and cellular fucosylation. Proc Natl Acad Sci USA 2013; 110(14): 5404-9.
[http://dx.doi.org/10.1073/pnas.1222263110] [PMID: 23493549]

[55] Li H, Sethuraman N, Stadheim TA, *et al*. Optimization of humanized IgGs in glycoengineered *Pichia pastoris*. Nat Biotechnol 2006; 24(2): 210-5.
[http://dx.doi.org/10.1038/nbt1178] [PMID: 16429149]

[56] Cox KM, Sterling JD, Regan JT, *et al*. Glycan optimization of a human monoclonal antibody in the aquatic plant Lemna minor. Nat Biotechnol 2006; 24(12): 1591-7.
[http://dx.doi.org/10.1038/nbt1260] [PMID: 17128273]

[57] Mabashi-Asazuma H, Kuo CW, Khoo KH, Jarvis DL. A novel baculovirus vector for the production of nonfucosylated recombinant glycoproteins in insect cells. Glycobiology 2014; 24(3): 325-40.
[http://dx.doi.org/10.1093/glycob/cwt161] [PMID: 24362443]

[58] Huang W, Giddens J, Fan SQ, Toonstra C, Wang LX. Chemoenzymatic glycoengineering of intact IgG antibodies for gain of functions. J Am Chem Soc 2012; 134(29): 12308-18.
[http://dx.doi.org/10.1021/ja3051266] [PMID: 22747414]

[59] Smith EL, Giddens JP, Iavarone AT, Godula K, Wang LX, Bertozzi CR. Chemoenzymatic Fc glycosylation *via* engineered aldehyde tags. Bioconjug Chem 2014; 25(4): 788-95.
[http://dx.doi.org/10.1021/bc500061s] [PMID: 24702330]

[60] Beck A, Reichert JM. Marketing approval of mogamulizumab: a triumph for glyco-engineering. MAbs 2012; 4(4): 419-25.
[http://dx.doi.org/10.4161/mabs.20996] [PMID: 22699226]

[61] Gagez AL, Cartron G. Obinutuzumab: a new class of anti-CD20 monoclonal antibody. Curr Opin Oncol 2014; 26(5): 484-91.

[http://dx.doi.org/10.1097/CCO.0000000000000107] [PMID: 25014645]

[62]   Herting F, Friess T, Bader S, *et al.* Enhanced anti-tumor activity of the glycoengineered type II CD20 antibody obinutuzumab (GA101) in combination with chemotherapy in xenograft models of human lymphoma. Leuk Lymphoma 2014; 55(9): 2151-5160.
[http://dx.doi.org/10.3109/10428194.2013.856008] [PMID: 24304419]

[63]   Mössner E, Brünker P, Moser S, *et al.* Increasing the efficacy of CD20 antibody therapy through the engineering of a new type II anti-CD20 antibody with enhanced direct and immune effector cell-mediated B-cell cytotoxicity. Blood 2010; 115(22): 4393-402.
[http://dx.doi.org/10.1182/blood-2009-06-225979] [PMID: 20194898]

[64]   Niederfellner G, Lammens A, Mundigl O, *et al.* Epitope characterization and crystal structure of GA101 provide insights into the molecular basis for type I/II distinction of CD20 antibodies. Blood 2011; 118(2): 358-67.
[http://dx.doi.org/10.1182/blood-2010-09-305847] [PMID: 21444918]

[65]   Illidge T, Cheadle EJ, Donaghy C, Honeychurch J. Update on obinutuzumab in the treatment of B-cell malignancies. Expert Opin Biol Ther 2014; 14(10): 1507-17.
[http://dx.doi.org/10.1517/14712598.2014.948414] [PMID: 25190612]

[66]   Anthony RM, Wermeling F, Karlsson MC, Ravetch JV. Identification of a receptor required for the anti-inflammatory activity of IVIG. Proc Natl Acad Sci USA 2008; 105(50): 19571-8.
[http://dx.doi.org/10.1073/pnas.0810163105] [PMID: 19036920]

[67]   Anthony RM, Kobayashi T, Wermeling F, Ravetch JV. Intravenous gammaglobulin suppresses inflammation through a novel T(H)2 pathway. Nature 2011; 475(7354): 110-3.
[http://dx.doi.org/10.1038/nature10134] [PMID: 21685887]

[68]   Schwab I, Mihai S, Seeling M, Kasperkiewicz M, Ludwig RJ, Nimmerjahn F. Broad requirement for terminal sialic acid residues and FcγRIIB for the preventive and therapeutic activity of intravenous immunoglobulins in vivo. Eur J Immunol 2014; 44(5): 1444-53.
[http://dx.doi.org/10.1002/eji.201344230] [PMID: 24505033]

[69]   Washburn N, Schwab I, Ortiz D, *et al.* Controlled tetra-Fc sialylation of IVIg results in a drug candidate with consistent enhanced anti-inflammatory activity. Proc Natl Acad Sci USA 2015; 112(11): E1297-306.
[http://dx.doi.org/10.1073/pnas.1422481112] [PMID: 25733881]

[70]   Yu X, Vasiljevic S, Mitchell DA, Crispin M, Scanlan CN. Dissecting the molecular mechanism of IVIg therapy: the interaction between serum IgG and DC-SIGN is independent of antibody glycoform or Fc domain. J Mol Biol 2013; 425(8): 1253-8.
[http://dx.doi.org/10.1016/j.jmb.2013.02.006] [PMID: 23416198]

[71]   Ahmed AA, Giddens J, Pincetic A, *et al.* Structural characterization of anti-inflammatory immunoglobulin G Fc proteins. J Mol Biol 2014; 426(18): 3166-79.
[http://dx.doi.org/10.1016/j.jmb.2014.07.006] [PMID: 25036289]

[72]   Nandakumar KS, Collin M, Happonen KE, *et al.* Dominant suppression of inflammation by glycan-hydrolyzed IgG. Proc Natl Acad Sci USA 2013; 110(25): 10252-7.
[http://dx.doi.org/10.1073/pnas.1301480110] [PMID: 23671108]

[73] Hinman LM, Hamann PR, Wallace R, Menendez AT, Durr FE, Upeslacis J. Preparation and characterization of monoclonal antibody conjugates of the calicheamicins: a novel and potent family of antitumor antibiotics. Cancer Res 1993; 53(14): 3336-42.
[PMID: 8324745]

[74] Hamann PR, Hinman LM, Beyer CF, *et al.* An anti-MUC1 antibody-calicheamicin conjugate for treatment of solid tumors. Choice of linker and overcoming drug resistance. Bioconjug Chem 2005; 16(2): 346-53.
[http://dx.doi.org/10.1021/bc049795f] [PMID: 15769088]

[75] Zuberbühler K, Casi G, Bernardes GJ, Neri D. Fucose-specific conjugation of hydrazide derivatives to a vascular-targeting monoclonal antibody in IgG format. Chem Commun (Camb) 2012; 48(56): 7100-2.
[http://dx.doi.org/10.1039/c2cc32412a] [PMID: 22684082]

[76] Zhou Q, Stefano JE, Manning C, *et al.* Site-specific antibody-drug conjugation through glycoengineering. Bioconjug Chem 2014; 25(3): 510-20.
[http://dx.doi.org/10.1021/bc400505q] [PMID: 24533768]

[77] Okeley NM, Toki BE, Zhang X, *et al.* Metabolic engineering of monoclonal antibody carbohydrates for antibody-drug conjugation. Bioconjug Chem 2013; 24(10): 1650-5.
[http://dx.doi.org/10.1021/bc4002695] [PMID: 24050213]

[78] Boeggeman E, Ramakrishnan B, Pasek M, *et al.* Site specific conjugation of fluoroprobes to the remodeled Fc N-glycans of monoclonal antibodies using mutant glycosyltransferases: application for cell surface antigen detection. Bioconjug Chem 2009; 20(6): 1228-36.
[http://dx.doi.org/10.1021/bc900103p] [PMID: 19425533]

[79] Qasba PK, Boeggeman E, Ramakrishnan B. Site-specific linking of biomolecules *via* glycan residues using glycosyltransferases. Biotechnol Prog 2008; 24(3): 520-6.
[http://dx.doi.org/10.1021/bp0704034] [PMID: 18426242]

[80] Zhu Z, Ramakrishnan B, Li J, *et al.* Site-specific antibody-drug conjugation through an engineered glycotransferase and a chemically reactive sugar. MAbs 2014; 6(5): 1190-200.
[http://dx.doi.org/10.4161/mabs.29889] [PMID: 25517304]

[81] Zeglis BM, Davis CB, Aggeler R, *et al.* Enzyme-mediated methodology for the site-specific radiolabeling of antibodies based on catalyst-free click chemistry. Bioconjug Chem 2013; 24(6): 1057-67.
[http://dx.doi.org/10.1021/bc400122c] [PMID: 23688208]

[82] Zeglis BM, Davis CB, Abdel-Atti D, *et al.* Chemoenzymatic strategy for the synthesis of site-specifically labeled immunoconjugates for multimodal PET and optical imaging. Bioconjug Chem 2014; 25(12): 2123-8.
[http://dx.doi.org/10.1021/bc500499h] [PMID: 25418333]

[83] van Geel R, Wijdeven MA, Heesbeen R, *et al.* Chemoenzymatic Conjugation of Toxic Payloads to the Globally Conserved N-Glycan of Native mAbs Provides Homogeneous and Highly Efficacious Antibody-Drug Conjugates. Bioconjug Chem 2015; 26(11): 2233-42.
[http://dx.doi.org/10.1021/acs.bioconjchem.5b00224] [PMID: 26061183]

CHAPTER 7

Engineering of Therapeutic Proteins through Hyperglycosylation

Natalia Ceaglio[*], **Marina Etcheverrigaray**, **Ricardo Kratje** and **Marcos Oggero**

UNL, CONICET, Cell Culture Laboratory, FBCB. Edificio FBCB-Ciudad Universitaria UNL, C.C. 242. (S3000ZAA) Santa Fe, Argentina

Abstract: One of the major challenges associated with the development of protein-based biotherapeutics lies in achieving persistent concentrations of the active molecule in circulation. Most human proteins are rapidly cleared from circulation, mainly by renal filtration, and so increasing their *in vivo* residence time to reach appropriate therapeutic doses has been a matter of extensive investigation. The majority of therapeutic proteins exhibit post-translational modifications (PTMs). Among them, N- and O-glycosylation are the most abundant and complex modifications that proteins can undergo, affecting diverse biological properties, including solubility, protease and thermal stability, antigenicity, immunogenicity, bioactivity and pharmacokinetics. Thus, glycosylation represents one of the most relevant attributes of many therapeutic proteins, defining their potency and effectiveness. Also, both size and charge of proteins are completely modified by the presence of glycans, so that manipulation of this PTM represents a valuable tool to alter the pharmacokinetics and pharmacodynamics of biotherapeutics. This chapter deals with different glycoengineering strategies developed with the aim of increasing the plasma half-life of proteins, as well as other properties. Specifically, engineering of proteins through the addition of new glycosyl moieties is addressed. A thorough description of the properties conferred to proteins by glycans is first presented, followed by a description of the strategies developed for the rational manipulation of glycosylation parameters to improve such properties. Different approaches to incorporate new N- and O-glycans

[*] **Corresponding author Natalia Ceaglio:** UNL, CONICET, Cell Culture Laboratory, FBCB. Edificio FBCB-Ciudad Universitaria UNL. C.C. 242. (S3000ZAA) Santa Fe. Argentina; Tel/Fax: +54 342 4552928; E-mail: nceaglio@fbcb.unl.edu.ar

Qun Zhou (Ed.)

into proteins are described and exemplified. Finally, the application of N- and O-hyperglycosylation engineering to an emblematic protein such as recombinant human IFN-α2b is presented.

Keywords: Antigenicity, Bioactivity, Biotherapeutics, CTP, Efficacy, Glycoengineering, Glycosylation prediction, hCG, hEPO, hFSH, hIFN-α2b, Immunogenicity, N-glycosylation, O-glycosylation, Peptide fusion, Pharmacokinetics, Plasma clearance, Site-directed mutagenesis, Solubility, Stability.

INTRODUCTION

In the last decades there has been an accelerated development of protein-based biotherapeutics which have revolutionized the treatment of numerous diseases due to their high specificity and affinity towards its clinical target. Unfortunately, most proteins do not behave as ideal drugs, since their efficacy is compromised by several intrinsic limitations. These include the low *in vivo* activity displayed due to their low stability and short serum half-life, which represent the main obstacles to overcome in order to achieve a successful application as therapeutics [1 - 3]. Thus, many approaches have been developed with the aim of improving the residence time of proteins in circulation to reach appropriate therapeutic doses, as well as to improve other physicochemical and pharmacological properties [4, 5].

The mechanisms involved in plasma clearance of proteins are diverse [6]; however, the most commonly employed route is renal filtration. The structural and physicochemical characteristics of the glomerular barrier are responsible, to a great extent, of the elimination mechanisms. Small molecules are rapidly filtrated through the numerous small 4-10 nm diameter pores, while negatively charged glycosaminoglycans form an anionic barrier which selectively prevents the passage of macromolecules according to their charge. Another site of metabolism of biologics is the liver, where hepatocytes participate in their metabolism through receptor-mediated membrane transport and lysosomal degradation. In addition, peripheral blood-mediated elimination by proteolysis contributes to a great extent to the catabolism of administered proteins [7]. Thus, the structure and size of a protein, represented by its hydrodynamic ratio, as well as its physicochemical properties constitute starting points for improvement of its plasma half-life.

The majority of proteins derived from eukaryotes undergo covalent modifications either during or after their ribosomal synthesis. This gives rise to the concept of co-translational and post-translational modifications (PTMs). Glycosylation is the most abundant and complex PTM among eukaryotic cells. It constitutes an enzymatic process in which oligosaccharide chains are covalently attached to the amide nitrogen of an asparagine side chain (N-glycosylation) or to oxygen of a hydroxyl group of serine or threonine residues (O-glycosylation), although other types of glycosylation have been described [8, 9]. More than 50% of human proteins are glycoproteins and it is estimated that 1-2% of genome includes genes related to glycan synthesis. Also, glycosylation represents one of the most relevant attributes of many therapeutic proteins, defining their potency and effectiveness [10, 11]. Both size and charge of proteins are completely modified by the presence of glycans, so that manipulation of this PTM represents a valuable tool to alter the pharmacokinetics and pharmacodynamics of biotherapeutics [12]. Thus, engineering of proteins through the addition of new glycosyl moieties has emerged as an area of great technological interest.

PROPERTIES CONFERRED TO PROTEINS BY GLYCANS

There are many theories about the roles of carbohydrates in different types of glycoconjugates, but none of them has demonstrated to be of universal application [13]. Oligosaccharides play a marked effect over diverse biological properties of the glycoproteins, including solubility, stability (regarding protease resistance and thermal degradation), antigenicity, immunogenicity, bioactivity, cell interaction, pharmacokinetics and plasma half-life [14].

Effects of Glycosylation on Solubility and Stability of Glycoproteins

Oligosaccharides often promote protein solubility and prevent protein aggregation. For example, complete removal of O-glycans of human granulocyte colony-stimulating factor (hG-CSF) resulted in a high level of autoaggregation, which led to its biological inactivation [15]. Besides, glycans could contribute to the increment of solubility of proteins through masking exposed hydrophobic amino acid residues.

In vivo glycoproteins stability generally depends of their resistance to protease inactivation. Glycans can protect glycoproteins from extracellular proteolytic enzymes attack, probably by covering cleavage sites. The presence of terminal sialic acids protected erythropoietin (EPO) [16], plasminogen tissue activator (tPA) [17] and interferons (IFNs) from proteolytic inactivation. Besides, in some cases, total removal of proteins oligosaccharides reduced their thermal stability. For instance, enzymatic hydrolysis of EPO sialic acids increased its denaturation rate at 70°C [14, 18, 19].

Effects of Glycosylation on Bioactivity of Glycoproteins

In many cases, carbohydrates exert a profound effect over the biological activity of a glycoprotein. Glycans contain the information to establish specific interactions between cells, proteins-glycoproteins and oligosaccharides-lectins. Also, they are essential for the correct processing, secretion, folding and maintenance of glycoproteins in their active biological conformation. However, the precise effect of carbohydrates on glycoprotein´s bioactivity cannot be generalized nor predicted. There are examples in which the absence of glycans causes an increment of the *in vitro* specific biological activity: aglycosylated prolactin exhibits a higher affinity for its receptor and a higher ability to stimulate cell proliferation than its glycosylated counterpart [20]. In other cases, the deficiency of oligosaccharides reduced the bioactivity of glycoproteins: for example, removal of sialic acids abolishes the coagulant activity of Factor IX [21]

For many proteins, complete glycosylation is equivalent to maximum specific biological activity, although this does not mean a great impact over *in vitro* activity. For example, although human desialylated EPO exhibits a 1000-fold reduction in its *in vivo* activity compared to native EPO, the increment in *in vitro* bioactivity is less pronounced [14, 18].

Effects of Glycosylation on Antigenicity and Immunogenicity

Glycosylation contributes to the antigenic properties of glycoproteins through different ways: by inactivating the peptide epitope (antigenic sites masking) or by constituting part of the structure recognized by the antibody. Some oligosaccharide chains can be antigenic themselves and generate anti-carbohydrates antibodies, while others need the protein context to be recognized [22].

The effects of glycans over the immunogenicity of glycoproteins, *i.e.*, the ability to stimulate an immune response, are less clear. No anti-EPO antibodies have been found in patients treated with the hormone produced in Chinese Hamster Ovary (CHO) cells, which has the same amino acid sequence as the natural protein but exhibits some differences in its glycosyl moiety. Besides, natural glycosylated proteins could be immunogenic if they are administered in their non-glycosylated form [14]. Indeed, the induction of antibodies in patient's serum during disease treatment using non-glycosylated interleukin-2 (IL-2), interferon-γ (IFN-γ), granulocyte and macrophage colony-stimulating factor (GM-CSF) and G-CSF has been reported [23, 24]. Also, natural antibodies were found in normal serum from healthy patients, which bound to non-glycosylated cytokines to a higher extent [25]. In many cases, the antibodies were able to neutralize the bioactivity of the mentioned cytokines. Consequently, the choice of an adequate cellular host for the production of glycosylated proteins would reduce the risks of development of a humoral immune response during their therapeutic use.

Effects of Glycosylation on Pharmacokinetics and *in vivo* Clearance

Oligosaccharides play a key role in defining the *in vivo* clearance rate of glycoproteins, which is an important property related to their therapeutic efficacy. The *in vitro* bioactivity would not be of major importance if the injected glycoprotein is rapidly cleared from circulation. For example, *in vitro* bioactivity of EPO increased after desialylation, but *in vivo* activity decreased considerably due to its fast plasma elimination [26].

As mentioned above, human proteins with molecular masses smaller than 70 kDa are continuously removed from plasma through glomerular filtration. The rate of filtration through renal tubules is sensitive to the tertiary structure of the protein and its molecular mass. In addition, filtration can be restricted through charge repulsion, since the glomerular barrier is negatively charged due to the presence of glycosaminoglycans. Thus, in low molecular mass glycoproteins, glycans could prolong plasma half-life through an increase of the molecular size and the negative charge provided by sialic acids [14, 27].

There are different mechanisms of circulatory clearance associated to high affinity receptors that interact with terminal monosaccharides of glycoproteins. The asialoglycoprotein receptor expressed in hepatocytes binds to glycoproteins which exhibit a terminal galactose or N-acetylgalactosamine, causing their plasma elimination [28, 29]. This mechanism is related to the clearance of glycoproteins which lack terminal sialic acids, and it is considered the main factor resulting in fast plasma clearance of desialylated EPO. Mannose receptor-mediated clearance of proteins with terminal mannose, such as those produced by yeast, has also been described [30].

Specific receptor-mediated elimination can also be a major effect on serum half-life of proteins. In this scenario, the lower binding affinity of glycosylated proteins than their aglycosylated counterparts may also contribute to reduce their clearance rate [31].

Moreover, glycans also play an important role in protecting from extracellular proteolytic enzymes through masking of cleavage sites. In this way, protein elimination from circulation through inactivation from serum proteases is reduced.

The biological roles of glycans in nature encompass the spectrum from those that are almost negligible to those that are crucial for the development, growth, function or survival of an organism. Particularly, they exert important effects over the glycan-bearing proteins, so that manipulation of glycan content and/or structure could be rationally used for modulating the activity of glycoproteins.

INCREASING THE EFFICACY OF RECOMBINANT PROTEIN-BASED THERAPIES THROUGH GLYCOSYLATION ENGINEERING

Glycosylation of therapeutic proteins is crucial for their biological activity. Protein glycosylation also plays a key role in numerous biological processes, affecting protein folding, protein stability and solubility as well as binding to other proteins such as receptors. Considering this, the rational manipulation of glycosylation parameters is widely applied to improve the above-mentioned properties and, consequently, the biological actions of those proteins. Therefore, glycoengineering arises as a strategy based on the modification the content and/or the structure of protein´s glycans through the adjustment of culture conditions or genetic modifications.

Thus, controlling or modifying different biosynthetic pathways are approaches by which glycoengineering intends to make proteins bearing more defined glycans. For example, many approaches were developed for expressing N-glycans with active conformations in prokaryotes [32 - 35]; in eukaryotes including yeast, insect cells or plants, efforts were directed to introduce the ability of producing complex-types of glycans and/or to avoid the synthesis of immunogenic N-glycans [36 - 40]. Also, in mammalian cells, and particularly in CHO cells, glycoengineering was carried out in order to make glycoproteins more homogeneous regarding the repertoire of glycoforms [41], to produce them into a context of a more human-like glycosylation [42] or to increase their bioactivities as it was extensively carried out in the case of antibodies [43, 44].

An interesting and new glycoengineering approach is the GlycoDelete strategy. It was firstly developed in the human HEK293 cell [45] and was then carried out in plant cells [46]. This system tries to solve some issues related with N-glycosylation heterogeneity during the production of mammalian glycoproteins and to avoid the immunogenicity that is frequently observed when plant-derived glycoproteins are used as biotherapeuctics (complex-type N- glycans from plants exhibit β-1,2-xylose and α-1,3-fucose that are highly immunogenic). By means of this technology, a modification in the glycosylation machinery is introduced by inactivating an N-acetylglucosaminyltransferase I (the MGAT1 gene-encoded GnTI) and by overexpressing a deglycosylating enzyme (endo-N-acetyl-β-D-glucosaminidase; endo T) targeted to the Golgi apparatus. This enzyme hydrolyzes the high-mannose type glycans, which are present in proteins expressed in GnTI knockout cells. Hence, these modifications render glycoproteins having a single GlcNAc in the corresponding Asn. In mammalian hosts, the GlcNAc residues can be further elongated by galactosyl and sialyltransferases to produce proteins with the Gal-GlcNAc disaccharide or its α-2,3-sialylated derivative and some monosacharide intermediates [45]. In the case of plant cells, the GlycoDelete glycoengineering strategy produces glycoproteins having single GlcNAc N-glycans where no further modifications of the GlcNAc to larger oligosaccharides are observed [46]. In this way, the Glycodelete is a procedure that allows the glycoproteins to retain the folding-enhancing actions that N-glycans exert during translation and to avoid their broad heterogeneity. Consequently, this strategy could be used to produce recombinant proteins in which their functions are not affected by the lack of complex-type N-glycans. Besides, it simplifies their purification process and lowers their immunogenic properties.

Nevertheless, glycoengineering not only strives for a procedure to optimize the glycosylation of a protein or a way to remodel existing glycan structures but it can also be defined as an approach for increasing *in vivo* activity and prolonging duration of action of proteins by introducing new sialic acid containing carbohydrates into them [47]. In this regard, Sinclair *et al.* [48] indicated that one aspect of glycoengineering is to introduce new N-linked glycosylation consensus sequences into desirable positions in the peptide backbone to generate proteins with longer serum half-life. Besides, Elliott *et al.* [47] showed that the glycoengineering strategy based in the introduction of new N-glycosylation sites could be an applicable method for increasing the half-life of proteins which already contain N-glycosylation sites (such as erythropoietin), proteins which normally contain only O-linked carbohydrates (such as MpI Ligand) and proteins which normally lack carbohydrates entirely (such as leptin). Moreover, the addition of O-glycans to a protein also matches the concept of "introducing new sialic acid containing carbohydrates onto the target protein" and validates the designation of glycoengineering.

Several human proteins were modified by adding new N- or O-glycosylation sites: hEPO [49 - 51], human follicle stimulating hormone (hFSH) [52 - 54], hIFN-α2b [55, 56] and human hIFN-β [57, 58], among others. With the aim of improving the pharmacokinetic properties and stability of some proteins and thus, conferring a greater *in vivo* potency, second-generation of biotherapeutics optimized by introducing new N- or O-glycosylation sites were obtained. Applying this glycoengineering strategy, two biopharmaceutics were approved in treating patients in clinics: darbepoietin alfa or Novel Erythropoiesis Stimulating Protein (NESP™) derived from hEPO [49, 50] and corifollitropin alfa or Elonva™ derived from hFSH [59]. In this chapter, the glycosylation engineering by introducing N-/O-glycosylation sites will be described and exemplified.

ENGINEERING OF PROTEINS THROUGH HYPERGLYCO-SYLATION

The introduction of new glycosylation can be exploited to modulate the bioactivity of proteins which already contain N- and/or O-linked oligosaccharides or to those which lack carbohydrates at all. Thus, engineering of proteins through the addition of new glycosylation sites represents a strategy for improving the *in vivo* efficacy of therapeutic proteins, particularly, by prolonging their plasma half-lives.

Increment of glycan content of a protein or glycoprotein can be done through the attachment of N-linked or O-linked carbohydrates. This could be performed through chemical conjugation of monosaccharides or oligosaccharides to the protein. Although this system has the advantage of being easily controlled, it requires additional steps during the production process [60]. Besides, many examples demonstrated the success of genetic modification of proteins to generate or add new recognition sites for glycan addition [47, 55, 61 - 63]. The main challenge of this approach lies in the rational selection of the adequate positions to incorporate glycosyl moieties, preserving at the same time the functional and structural properties of the protein of interest [1].

Engineering of Proteins through the Addition of N-glycans

Factors Affecting N-glycosylation Efficiency

N-glycosylation begins with the synthesis of a precursor on a lipid anchor (dolicol) through the incorporation of monosaccharides by a set of glycosyltransferases. Oligosaccharyltransferase (OST), the central enzyme of this pathway, catalyzes the transference of the lipid-linked oligosaccharide to the consensus sequence Asn-Xxx-Ser/Thr of the nascent protein in the endoplasmic reticulum (ER). The oligosaccharide is further extended and processed in the ER and Golgi apparatus through a set of enzymatic reactions in order to yield the different oligosaccharide structures [64, 65] (Fig. **1**).

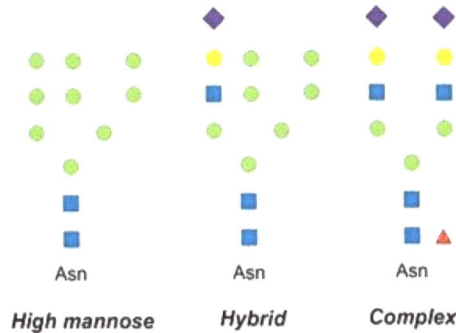

Fig. (1). Graphical representation of N-glycan oligosaccharide structures. High mannose, hybrid and complex oligosaccharides share the pentasaccharide core Man$_3$GlcNAc$_2$. Symbols: ■ N-Acetylglucosamine (GlcNAc), ● Mannose (Man), ● Galactose (Gal), ▲ Fucose (Fuc), ◆ N-Acetylneuraminic acid (Neu5Ac).

However, although all translated and secreted proteins in a cell are exposed to the same glycosylation machinery, not all potential N-glycosylation sites are invariably occupied. In contrast to polypeptide synthesis, which proceeds under strict genetic control, oligosaccharides are attached to the consensus sequence N-X-S/T in the absence of nucleic acid guidance. Thus, the sole presence of the consensus sequence in a protein does not guarantee the existence of N-linked oligosaccharides. A single polypeptide that is glycosylated normally emerges from the biosynthetic pathway as a mixture of glycosylation variants, known as glycoforms. The frequency of glycosylation on a given site is variable and determines a macroheterogeneity of glycoforms, *i.e.*, specific glycosylation sites on a protein may or may not carry a glycan. Besides, a great variety of glycan structures may be attached to the same site, giving rise to the concept of microheterogeneity of glycoforms [66, 67].

It is estimated that between 10-30% of potential N-glycosylation sites are not occupied. The first challenge in glycosylation engineering is to achieve maximum site occupancy, since it is not completely understood under which circumstances certain glycosylation sites are skipped or occupied. The glycosylation efficiency depends on the availability of precursors linked to the lipid dolicol, the OST activity and the primary, secondary and tertiary structure of a protein [68].

Site occupancy is highly influenced by the primary sequence of the protein, particularly, by the amino acids that are near to or forming part of the consensus tripeptide Asn-Xxx-Ser/Thr. The structure of the central residue (Xxx) of the tripeptide is an important determinant of the glycosylation eficiency. Site-directed mutagenesis studies have demonstrated that Pro completely blocked N-glycosylation, while sequences containing Trp, Asp, Glu and Leu were inefficiently glycosylated [69]. This could be related to the physicochemical characteristics of amino acid residues. Large hydrophobic amino acids may impair oligosaccharide transfer, while the negative charge of some residues could increase the interaction of OST. Also, the presence of hydroxyl groups were correlated to a higher glycosylation efficiency. Another factor that completely inhibits glycosylation is the presence of Pro immediately after the consensus site. In addition, there seems to be a preference for Asn-Xxx-Ser/Thr sites over Asn-Xxx-Ser/Thr, resulting in an occupation degree between 2 to 3 times higher, which correlates with the substrate specificity of OST [64, 65, 70]. However, recent findings have given evidence of less strict requirements for the acceptor site in the glycoprotein. For example, glycosylation has been reported in asparagine residues of non-consensus sequences in different recombinant antibodies produced in CHO cells [71]. Regarding the position of the N-glycosylation site, it has been reported that consensus sequences near both ends of the polypeptide were less efficiently glycosylated [72, 73]. Particularly, it has been suggested that carbohydrate addition does not occur at C-terminal positions because of the reduced amount of time available to add carbohydrate to an almost completely synthesized protein [74].

Regarding secondary structure, studies have revealed that 70% of consensus sequences are located in β-bends, 20% in β-sheets and 10% in α-helices [66]. Disordered-regions like loops would be more favorable regions for addition of N-glycans and could enhance the stability of the glycoprotein, in contrast with structured and less flexible regions such as helices [75]. Although normally glycosylation efficiency is not influenced by the 3D structure of the protein, the structural configuration of the nascent polypeptide could influence the accessibility of the tripeptide to the action of the OST. Since N-glycosylation is a co-translational process, the glycosylation efficiency results from a competition between the rate of folding and the rate of addition of the lipid-linked precursor

[66, 76].

Macro- and microheterogeneity of glycoforms are also highly influenced by the host cell type, its physiological state and external conditions [77].

The presence of N-glycosylation consensus sequences is then necessary but not sufficient to ensure carbohydrate addition, and so other requirements may be fulfilled for a successful glycoengineering strategy development.

Approaches to Incorporate New N-glycan Moieties into Proteins

Introduction of Point Mutations to Generate New N-glycosylation Sites

A successful approach to generate hyper-N-glycosylated proteins is to introduce new Asn-Xxx-Ser/Thr consensus sequences into the protein sequence by site-directed mutagenesis of its cDNA and its subsequent expression in mammalian cells. However, modifications of amino acids should be carefully selected, in order to achieve the two main goals of this technology: first, to generate consensus sites with a high probability of glycosylation, and second, to avoid altering the 3D-structure and stability of the protein, with the aim of preserving its biological activity.

Although many amino acid residues can be mutated in order to generate the new consensus sequences, the most conservative approach is to modify the least number of residues as possible [55, 57]. In this regard, taking into account the consensus tripeptide for N-glycosylation (Asn-Xxx-Ser/Thr) two paths can be taken:

- Localization of every residue of the protein sequence which has a Ser and/or Thr residue in position +2 and exchange for Asn.
- Localization of every Asn residue of the protein sequence and mutations of the residue in position +2 by Thr, since, as it has been mentioned before, the presence of Thr in the third position of the consensus sequence results in an occupation degree two or three times higher than if the same position is occupied by Ser [70].

Once such sites have been identified, theoretical analysis in terms of glycosylation probability and preservation of the protein's function should be performed.

Considering factors affecting glycosylation efficiency, muteins containing consensus sites near the C-terminal end of the protein should not be considered appropriate candidates. Also, muteins containing Pro in the second position or immediately after the consensus site are predicted to be devoid of N-glycosylation. Due to their size and hydrophobicity, amino acid residues such as Leu, Glu, Asp and Trp may not be suitable to occupy the second position of the consensus tripeptide. The probability of glycosylation can also be analysed by many online N-glycosylation predictor servers, such as NetNGlyc 1.0 Server [78], GlycoEP Server [79] and NGlycPred [80], among others.

From another point of view, point mutations should not cause a significant distortion of the tertiary structure of the protein, which leads to loss of protein's bioactivity. In this regard, the first residue of the consensus sequence (Asn or the amino acid to be mutated to Asn) should exhibit a high solvent accessibility. This parameter can be calculated from online software such as ASAview [81] and renders information about the exposure degree of the residue to which the glycan chain would be attached. Since N-glycosylation is a co-translational process, if such residue is buried in the 3D structure of the protein, its mutation or the addition of an oligosaccharide chain may alter the correct folding of the protein, leading to a potential loss of biological activity. Moreover, in the case of cytokines, hormones or any protein whose activity is based on ligand/receptor/substrate interaction, studies to identify residues presumably involved in such interaction should be performed.

An outstanding example of the introduction of point mutations to generate new glycosylation sites is the development of NESP™ (Novel Erythropoiesis Stimulating Protein), a hyperglycosylated form of rhEPO which contains two additional N-glycosylation sites. For more than 20 years, rhEPO has been used for the treatment of anemia associated with chronic renal failure, cancer, and HIV infection. However, a major drawback of this therapy is that the recommended dose frequency is three times per week by subcutaneous or intravenous injection, and the duration of therapy is, in most cases, for the life of the patient. rhEPO is composed by a single 165-amino acid polypeptide which contains three N-linked glycan chains attached to Asn24, Asn38 and Asn83, and an O-glycosylation site in Ser126. It has been largely demonstrated that glycans are crucial determinants of rhEPO *in vivo* bioactivity. Although the protein with glycoforms containing low sialic acid content exhibits high affinity for the EPO receptor, and thus, high *in vitro* activity, it is rapidly cleared from plasma. In contrast, the protein with glycoforms containing the highest sialic acid content demonstrates an extended half-life and, consequently, it comprises the most active isoforms *in vivo*. These findings have led to the conclusion that plasma clearance (and not receptor binding) is the primary determinant of rhEPO *in vivo* biological activity, and that the amount of sialic acid present in rhEPO´s glycans regulate plasma clearance [49].

In an attempt to corroborate this hypothesis and to use this knowledge to generate a rhEPO analogue with increased *in vivo* bioactivity, and thus, reduced frequency of administration, darbopoietin alfa or NESP was developed [49, 50]. Using an N-glycoengineering strategy, rhEPO analogues with new N-glycosylation consensus sequences introduced by point mutations were screened for glycosylation degree, retention of *in vitro* activity, receptor binding and structural stability through a conformational immunoassay [47]. Then, the most suitable consensus sequences were combined to construct analogues with multiple N-glycosylation sites, resulting in the development of NESP, which contains two additional N-linked carbohydrate chains introduced by changing five amino acid residues in rhEPO using site-directed mutagenesis. Compared to rhEPO, darbopoietin alpha exhibits a three-fold longer half-life, leading to a 13-fold greater *in vivo* potency (determined as its ability to increase hematocrit in healthy mice) and it can be administered less frequently to obtain an equivalent biologic response. Importantly, NESP is the most effective rhEPO variant for the treatment of patients with anemia associated to renal failure that has been approved until this moment.

Fusion to Peptides Containing N-glycosylation Sites

An alternative pathway to introduce new N-glycosylation sites lies in the addition of small peptides encompassing such sites to the terminal ends of the protein. As it is known that the frequency of N-glycosylation is reduced when the consensus sites approach the C-terminal end of the protein [73], N-terminal elongations are generally preferred. In this way, mutations of amino acids forming part of the primary structure of the molecule are avoided, probably minimizing the distortion of the 3D structure. Moreover, many N-glycosylation sites could be potentially introduced through the attachment of a sole N-terminal elongation, skipping the tedious task of selecting the adequate positions to mutate in order not to interfere with receptor binding nor disrupt protein tertiary structure. However, this methodology is limited to proteins whose N-terminal end is not involved in receptor binding or biological function.

This methodology has been successfully applied to hFSH [52]. Three different N-terminal extensions comprising four to nine amino acids which included one or two consensus tripeptides for N-glycosylation were evaluated: ANIT, ANITV and ANITVNITV. These peptides included relatively small, nonpolar amino acid residues (Ile, Val and Ala) with the aim of minimizing steric hindrance. In general, extension muteins were glycosylated to a higher degree than FSH muteins containing single point mutations to generate Asn-Xxx-Ser/Thr sequences, indicating the success of the applied methodology. Particularly, the extension mutein containing two additional N-glycosylation sites exhibited a 3 to 4-fold longer half-life than native FSH. Moreover, although the specific activity of the extension variant was only 25% of the corresponding one of wild-type FSH, it exhibited a markedly increased *in vivo* potency (determined as its ability to increase ovarian weight in rats), making this protein a potential candidate for infertility treatment.

Engineering of Proteins through the Incorporation of O-glycans

General Aspects of O-glycosylation in Eukaryotes

The most abundant forms of glycosydic modifications on secreted and membrane-bound proteins are N-glycosylation and the mucin-type O-glycosylation. The latter consists in an evolutionarily conserved protein modification among eukaryotes: from unicellular organism as protozoa to multicellular ones as mammals [82]. In general, O-glycosylation consists in an N-acetylgalactosamine (GalNAc) residue linked to Ser or Thr residues (mucin-type), but other types of less abundant non-mucin O-linked glycans have been described. Regarding the attachment site, O-glycans can also be attached to Tyr [83 - 85] and hydroxylysine (Hyl) [85 - 87]. In relation to the amino acid-linked monosaccharide, O-glycosylation initiated by O-Man, O-Fuc, O-Gal, O-Xyl, O-Gal and O-GlcNAc has been found.

The addition of carbohydrates to a peptide backbone to generate a mucin-type O-glycoconjugate (henceforth simply referred to as O-glycosylation) is a posttranslational event that occurs through a covalent modification of the amino acids Ser or Thr [66, 88]. Differently from N-glycosylation, the consensus sequence corresponding to O-glycosylation is unknown, but evidences show a preference for Ser/Thr/Pro-rich domains, where Thr is the residue with the highest probability to be glycosylated [66, 88]. In particular, the presence of Pro in the nearness of the O-glycosyl supporting amino acids is related to the β-turn conformation of Pro in the polypeptide chain. It is feasibly that the role of such residue is to confer a proper Ser and Thr exposition that leads to more efficient O-glycosylation.

In mammals, the O-glycosylation is initiated in the cis-Golgi by up to twenty GalNAc-transferases (GalNAc-Ts) to form the known Tn antigen. They are a family of enzymes named as UDP-N-acetylgalactosamine: polypeptide N-acetylgalactosaminyltransferases (ppGaNTases, EC 2.4.1.41) that transfer GalNAc from the sugar donor UDP-GalNAc to Ser or Thr residues to form the alpha anomeric linkage GalNAcα1-O-Ser/Thr in a recently synthesized polypeptide chain, *i.e.*, after most protein folding events have taken place [85]. This large number of polypeptide GalNAc-T isoenzymes, which not only begin the O-glycan synthesis but also control it, is distinctive for this sort of protein glycosylation. The isoenzymes display tissue-specific expression in adult mammals and show several substrate specificities [89]. In both senses, Steentoft *et al.*, 2013 [90] provided the first evidence that the O-glycoproteome is differentially regulated in different cell types. Also, based on *in vitro* analysis of peptide substrate specificities of a large number of GalNAc-T isoforms, Kong *et al.* suggested that GalNAc-T1-T3 are the main isoenzymes that contribute to generate the O-glycoproteome in most cell lines, while others (GalNAc-T4, T5, T7, T11, T12, T14 andT16) seem to have a lower contribution with overlapped functions [91].

Once the GalNAc residues are linked to Ser or Thr, they are further processed through the addition of different monosaccharides (Gal, GlcNAc, sialic acid and fucose) by the action of more than 30 glycosyltransferases to form linear or branched O-GalNAc glycans [85]. Specifically, the Tn structure is sialylated or elongated to give rise to 8 cores that can be afterwards extended, 4 of which are the most commonly found (Fig. **2**) [87]. O-glycans are generally short, while N-glycans are larger with several branches that determine their high antennarity and thus, the high number of glycoforms.

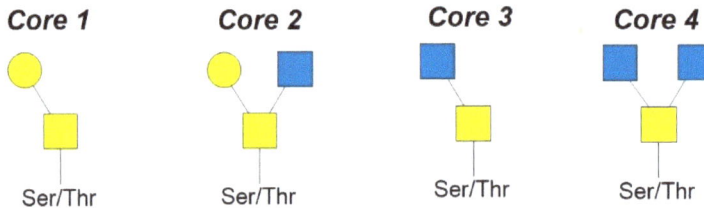

Fig. (2). Schematic representation of the most common mucin-type O-glycan cores produced by mammalian cells. Cores are then extended by addition of other monosaccharide residues, including Neu5Ac for termination.

So far, the information related to structural properties of mucin domains relies on its stiffened and extended conformation with no preference for the globular forms [92, 93]. There are also clear evidences of O-glycan location in disordered regions, while N-glycans in the same subset of proteins are mainly located in folded domains [90]. Accordingly, it has been demonstrated that O-glycosydic linkages were preferably placed in coil, turn or linker regions connecting domains and near to either the N- or C-terminal ends of proteins [93].

Mucin-type glycans exist in the form of eight distinct core structures where GalNAc is the nucleus of all them. In humans, the most common glycans bound to proteins are those corresponding to core 1 (Galβ1–3GalNAc) and core 2 (Galβ1–3(GlcNAcβ1–6)GalNAc). Core 1 is represented for small glycans ended with sialic acid while core 2 glycans tend to be built by larger glycans [82, 94].

Strategies to Introduce Potential O-glycosylation Sites

Both the absence of a consensus sequence for O-glycosylation and its structural features determine that protein glycoengineering by introducing O-glycosylation sites might be carried out by seeking for peptide domains from natural proteins that contain a dense concentration of O-glycans. Therefore, fusing peptides containing potential O-glycosylation sites to the N- and/or C-terminal ends of proteins represents a possibility to address this issue, avoiding the extensive modification of the peptide backbone [56]. For example, a well-known peptide with such characteristics is the carboxyl-terminal peptide (referred to as CTP) derived from the β-subunit of the human chorionic gonadotropin (hCG) [95].

After selecting the sequence that is naturally O-glycosylated or designing a new target sequence containing a Ser/Thr/Pro-rich domain, it is important to analyze the O-glycosylation probability by bioinformatics. Several O-glycosylation prediction methods have been developed [96]. For example, the following servers are used: NetOglyc [93], CKSAAP_OGlySite [97] and ISOGlyP [98]. All of them are based on neural networks that predict mucin-type O-glycosylation sites in mammalian proteins.

Considering the large number of ppGalNAc T isoenzymes that also have different properties, the mucin-type O-glycosylation is more complex than the mechanism that govern the N-glycosylation [98]. Therefore, it is important to keep in mind that the majority of O-glycosylation predictors do not have taken into account the wide range of Ser/Thr/Pro combination in a potential O-glycosylation site as well as the peptide substrate specificities of the various ppGalNAc T isoforms. In this sense, the predictor ISOGlyP is highly recommendable because it allows evaluating the potential O-glycosylation site in terms of ppGalNAc T isoform specificity and gives the possibility to quantitatively differentiate the amino acid residue preferences (so-called enhancement values, EVP) of the catalytic domain of the ppGalNAc-Ts. Thus, EVPs greater than 1 indicate an increased preference for a given site to be glycosylated by the transferases, while values less than 1 would suggest a disfavored glycosylating activity [98].

The Carboxyl-terminal Peptide, CTP

The carboxyl-terminal peptide (CTP) is an exclusively heavily O-glycosylated carboxyl-terminal extension of the hCG, rich in Ser and Pro residues, which confers immunological and biological specificity to the hormone [99]. The hGC, together with pituitary gonadotropins (human luteinizing hormone, hLH and hFSH) and human thyroid stimulating hormone (hTSH) are a family of heterodimeric glycoprotein hormones that contain a common and identical α-subunit covalently bound to a β-subunit which determines the biologic specificity of each protein [95]. The β-subunit of hCG is distinguished from the others because of the presence of the CTP where 4 Ser residues are responsible of the existence of O-linked glycans [100]. This extension is believed to play a role in maintaining the prolonged half-life of hCG compared to the other hormones [101].

The technology of fusing one or more units of CTP to the N- and/or C-terminal ends of several proteins has been employed to improve their therapeutic properties. Examples of them are: β-subunit of hFSH [54, 102], hTSH [103], α-subunit of hEPO [104], human growth hormone (hGH) [105], hEPO [51, 106], hIFN-α2b [56], among others. The works carried out by fusing CTP to hFSH, hTSH, hCG and hEPO demonstrated that the heavily O-glycosylated peptide did not affect assembly, secretion, receptor binding affinity and/or *in vitro* bioactivity. Contrarily, Ceaglio *et al.* showed that the addition of CTP also accounted for the reduction of *in vitro* antiproliferative and antiviral biological activities, although none of them were completely abolished. Nevertheless, the addition of O-linked oligosaccharides by CTP fusion significantly increased the half-life and *in vivo* potency of such protein [56].

There is not much information about the glycan structures that are exposed in the CTP extension. So far, the two natural occurring variants of hCG, whose structural differences strictly lay in their carbohydrate structure, has been thoroughly described. In this regard, significant differences have been found between the 4 O-linked oligosaccharides attached to CTP: while a trisaccharide corresponding to monosialylated core 1 was the principal structure found in pregnancy hCG, an hexasaccharide consisting of bisialylated core 2 represented the highest percentage of glycans described in the hyperglycosylated hCG from choriocarcinoma patients [100, 107]. Furthermore, a double CTP-modified IFN-α2b produced in CHO cells resembled more the glycan structures found in normal cells rather than in transformed ones [56].

A therapeutic protein generated by the CTP technology, which is used in clinics, is corifollitropin alfa (Elonva™) [59]. This protein is a fusion product of hFSH and CTP [54] which was designed as a sustained ovarian follicle stimulant, mimicking the activity of native hFSH but having a long lasting activity. Corifollitropin alfa has been approved in more than 50 markets outside the United States, including the European Union, for women undergoing controlled ovarian stimulation with a gonadotropin releasing hormone (GnRH) antagonist-assisted reproduction protocol [59]. Croxtall and McKeage emphasized that a treatment of seven once-daily injections of rhFSH was equivalent to a single injection of corifollitropin alfa to achieve the same efficacy, conferring a well-tolerated and more convenient therapeutic option to induce multiple follicular growth prior to assisted reproduction [59].

Therefore, it has been shown that CTP fusion represents a proper technology to augment the *in vivo* potency of therapeutic proteins, which have unfavorable pharmacokinetic properties. The origin and nature of CTP has made it attractive for the purpose of improving the pharmacokinetic properties of the selected proteins because CTP-fused proteins are unlikely to be toxic or immunogenic in humans.

HYPERGLYCOSYLATION OF HIFN-A2B AS AN EMBLEMATIC CASE

The naturally-O-glycosylated hIFN-α2b constitutes an example of a molecule that has been modified by introducing new N- or O-glycosylation sites in order to produce a novel biotherapeutic which displays the extensive benefits of glycan incorporation.

hIFN-α2b forms part of a group of pleiotropic cytokines (IFNs and specifically IFN-αs) that were first defined by their property of inducing paracrine resistance against virus infection [108], and then described by their potent antiproliferative and immunomodulatory effects [109].

Recombinant *E. coli*-derived hIFN-α2 (rhIFN-α2) is a widely used drug for the treatment of chronic hepatitis B and C [110], AIDS-related Kaposi´s sarcoma [111] and several types of cancer, including chronic myelogenous leukemia and hairy cell leukemia [112]. However, monotherapy with the bacteria-produced rhIFN-α2 requires daily or thrice-weekly injections into patients due to its short *in vivo* half-life (4–8 h in humans). Several different mechanisms, including rapid renal clearance, strong binding to specific receptors and proteolytic degradation, are responsible of its rapid clearance from blood. Consequently, the frequent administration of the cytokine leads to patient inconvenience and side effects, including skin, neurological, endocrine and immune toxicities [113].

Many strategies have been employed to improve the pharmacokinetic properties of the cytokine, aiming to increase its *in vivo* efficacy and thus, to confer a superior patient's life quality. One successful approach for improving the pharmacokinetics of rhIFN-α2b was the use of a glycoengineering strategy to obtain analogs with newly introduced potential N- or O-glycosylation sites [55, 56, 114, 115]. Therefore, IFN-α2b represents an emblematic case for this chapter.

N-glycoengineering of hIFN-α2b

Considering the amino acid sequence of hIFN-α2b, Ceaglio *et al.* designed twenty-eight potential N-glycosylation sites which were subjected to a theoretical analysis in terms of glycosylation probability and preservation of the protein's function (Table **1**) [55]. Fourteen mutations were then selected to be produced and thoroughly characterized.

Muteins containing 1-N-glycosylation site were transiently produced. Those having a high degree of glycosylation while retaining the highest *in vitro* biological activity were selected in order to generate IFN analogs owning from 2 to 5 N-glycosylation sites.

Muteins with multiple N-linked carbohydrates were assembled, expressed in CHO cells and analyzed in terms of glycosylation extent and *in vitro* bioactivity. The successive addition of a functional N-linked consensus sequence correlated with the increment of the analogs´ size. This could be detected in western blot analysis as bands of molecular masses higher than that corresponding to the wild type O-glycosylated hIFN-α2b (Fig. **3A**). Muteins having 3N, 4N and 5N glycosylation sites showed a great microheterogeneity of glycoforms, observed as a broad band between 27 and 44 kDa. Also, the charge-based heterogeneity of the IFN variants was analyzed by isoelectric focusing (IEF), demonstrating that each successive addition of a functional glycosylation site resulted in the appearance of a new series of isoforms with a lower pI (Fig. **3B**).

Table 1. Theoretical analysis of potential N-glycosylation sites to produce 1N-glycosylated IFN analogs. (Reprinted from Ceaglio *et al.* (2008) [55] with permission from Elsevier).

Mutein[a]	*Asn or Asn-substituted Residue[b]*	*Probability of Glycosylation (%)*	*ASA[d] (%)*	*Pro in Position +1 or +3*	*Interaction with the Receptor*
P4N	P4	57	34.5	No	No
T6N	T6	59	81.5	No	No
L9N	L9	64	32.8	No	No
R12N	R12	69	39.7	No	**Probable**
R23N	R23	64	48.6	No	No
L26N	L26	62	**21.8**	No	**Yes**
F47T	N45	58	83.3	No	No
A50N	A50	57	**18.1**	No	No
L66N	L66	45	**7.6**	No	No
F67T	N65	72	58.7	No	No
F67N	F67	71	**0.0**	No	No
K70N	K70	55	74.4	No	No
D71N	D71	**39[c]**	39.5	No	No
D77N	D77	60	47.8	No	No
F84N	F84	59	**0.0**	No	No
L95T	N93	69	62.2	No	No
G104N	G104	57	**7.6**	No	No
T106N	T106	**11**	57.7	**Yes**	No
E113N	E113	46	55.5	No	No
R125N	R125	59	51.1	No	No
K134N	K134	**12**	81.7	**Yes**	No
M148N	M148	**36**	24.5	No	**Yes**
S150N	S150	49	**0.0**	No	**Yes**

(Table 1) contd.....

Mutein[a]	Asn or Asn-substituted Residue[b]	Probability of Glycosylation (%)	ASA[d] (%)	Pro in Position +1 or +3	Interaction with the Receptor
S152N	S152	**36**	**26.4**	No	**Yes**
L153N	L153	52	**16.9**	No	**Yes**
Q158T	N156	60	39.6	No	No
Q158N	Q158	46	43.1	No	No
L161N	L161	46	45.9	No	No

[a]Muteins were designed as follows: original amino acid in the wild-type protein, position in the polypeptide chain, substituted amino acid in the mutated variant.

[b]The Asn or Asn-substituted residue likely to be glycosylated is indicated.

[c]Criteria which determined rejection of a mutein's construction are highlighted.

[d]Accesible surface area.

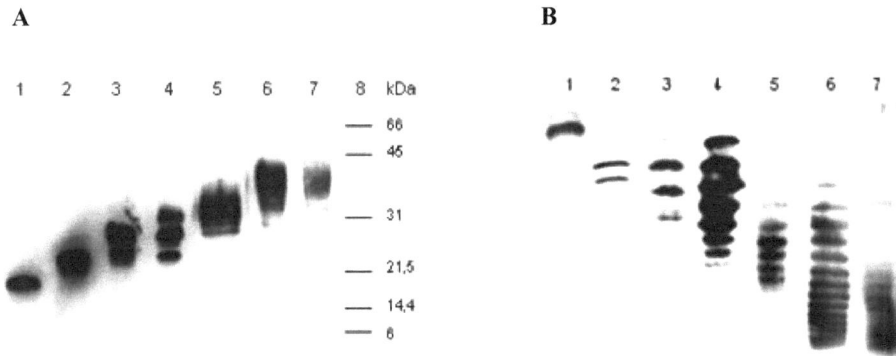

Fig. (3). Glycosylation patterns of N-glycosylated rhIFN-α2b analogs analyzed by SDS-PAGE followed by western blot (A) and by isoelectric focusing (B). Lane 1, non-glycosylated *E. coli*-derived rhIFN-α2b; lane 2, wild-type rhIFN-α2b; lanes 3-7, rhIFNa2b analogs with an increasing number of N-glycosylation sites (from 1 to 5); lane 8, molecular mass standards (only for SDS-PAGE assay). Reprinted from Ceaglio *et al.* (2008) [55] with permission from Elsevier.

In vitro specific antiviral bioactivity of IFN variants was gradually reduced with the incorporation of new glycosylation sites, showing a 10-fold lower potency than the corresponding to the wild-type cytokine, while 1% residual specific antiproliferative bioactivity was retained.

Pharmacokinetic studies in rats showed that all IFN analogs demonstrated better pharmacokinetic profiles than the wild-type and the non-glycosylated protein (Fig. 4). In particular, 4N and 5N analogs showed a 25-fold increased half-life and a

20-fold decreased apparent clearance regarding to the *E. coli*-derived cytokine. In accordance with their similar glycosylation profile, no significant differences between them were observed.

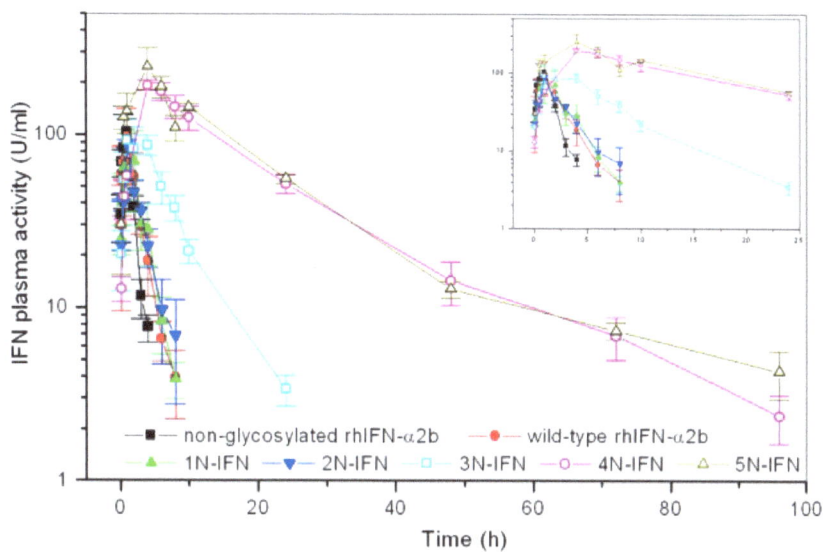

Fig. (4). Pharmacokinetic profiles of N-glycosylated rhIFN-α2b derivatives following subcutaneous injection in rats. Blood was collected at the time points indicated and assayed for IFN antiviral activity. Inset shows the data between 0 and 25 h. Reprinted from Ceaglio *et al.* (2008) [55] with permission from Elsevier.

The ability of IFN4N to inhibit the growth of solid tumors of prostate carcinoma-derived cells implanted in athymic nude mice was studied as a way to determine its *in vivo* efficacy. In comparison with the group treated with the non-glycosylated variant, a noteworthy reduction of the growth rate of the tumors at the end of the treatment with IFN4N was observed (Fig. **5**). Therefore, despite the significant reduction of the *in vitro* antiproliferative activity of the IFN4N analog, the increased molecular size and charge conferred by the new incorporated N-glycans determined an improved pharmacokinetic profile which was in turn responsible of its higher *in vivo* activity.

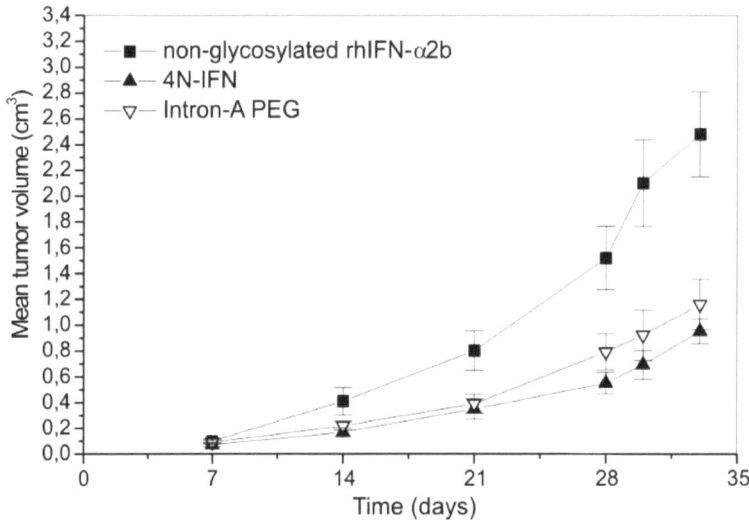

Fig. (5). *In vivo* antitumor activity of non-glycosylated rhIFN-α2b and 4N-IFN in athymic nude mice subcutaneously implanted with human prostate carcinoma PC-3 cells. Mice (n = 7) were injected s.c. with PC-3 cells and after two days they were treated with one weekly injection of each cytokine variant. The mean tumor volume ± standard error of the mean over time is shown. Reprinted from Ceaglio *et al.* (2010) [115] with permission from Elsevier.

O-glycoengineering of hIFN-α2b

The same authors applied an O-glycosylation engineering strategy to hIFN-α2b using the CTP fusing technology. For this, the CTP was fused to the N-terminal end (CTP-IFN), the C-terminal end (IFN-CTP) and both ends (CTP-IFN-CTP) of the cytokine [56].

To evaluate the O-glycosylation probability of CTP-fused variants, the ISOGlyP prediction server (http://isoglyp.utep.edu) was used [98]. Considering that O-GalNAc-Ts (T1–T3) are the most representative and active transferases [90] and also that the EVP was superior to 1 [91], 5 O-glycosylation sites were predicted to be occupied in proteins bearing a single CTP and 9 sites for CTP-IFN-CTP.

Stable CHO cell lines producing the three CTP-IFN variants were generated by lentiviral particles transduction. Then, all muteins were produced and purified from supernatants by immunoaffinity chromatography. IFN chimeras containing a single CTP exhibited different electrophoretic profiles, migrating with molecular masses of 37 kDa for the CTP-IFN-CTP, 31 kDa for CTP-IFN and 29 kDa for

IFN-CTP. All of them demonstrated higher molecular masses than the wild-type cytokine (21.5 kDa) (Fig. **6A**). Also, the mutein having 2 CTP fused to both ends of IFN showed a wide diversity of glycoforms (more than 10 bands), with a homogeneous profile since all of them were concentrated in the more acidic part of the pH range. Differently, the isoforms of the singly CTP-fused molecules were distributed all along the pH separation range (Fig. **6B**). The most acid isoform composition of CTP-IFN-CTP was consistent with a highest sialic acid content conferred by the new O-glycosyl moieties.

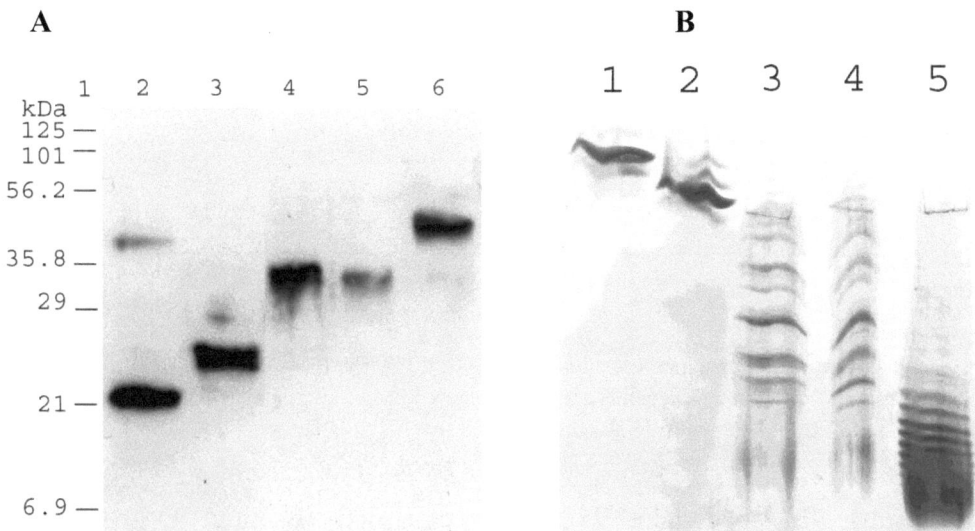

Fig. (6). (A) SDS-PAGE followed by western blot of CHO-K1 derived culture supernatants containing CTP-IFN chimeras. Lane 1, molecular mass standards; lane 2, non-glycosylated IFN; lane 3, wild-type IFN; lane 4, CTP-IFN; lane 5, IFN-CTP; lane 6, CTP-IFN-CTP. **(B)** IEF patterns of purified CTP-modified IFN analogs. Lane 1, non-glycosylated IFN; lane 2, wild-type IFN; lane 3, CTP-IFN; lane 4, IFN-CTP; lane 5, CTP-IF--CTP. Reprinted from Ceaglio *et al.* (2016) [56] with permission from Elsevier.

The incorporation of CTP reduced the *in vitro* specific antiproliferative activity of purified IFN variants which retained 6% (CTP-IFN), 8% (IFN-CTP) and 3% (CTP-IFN-CTP) of the corresponding activity of the wild-type rhIFN-α2b. Considering that the doubly CTP-modified IFN variant showed a 10-fold longer elimination half-life and a 19-fold decreased plasma apparent clearance compared to the wild-type cytokine Fig. (**7**) and that these parameters were quite similar to those corresponding to the IFN4N mutein, it is expected that CTP-IFN-CTP might develop a high *in vivo* efficacy (ongoing assays).

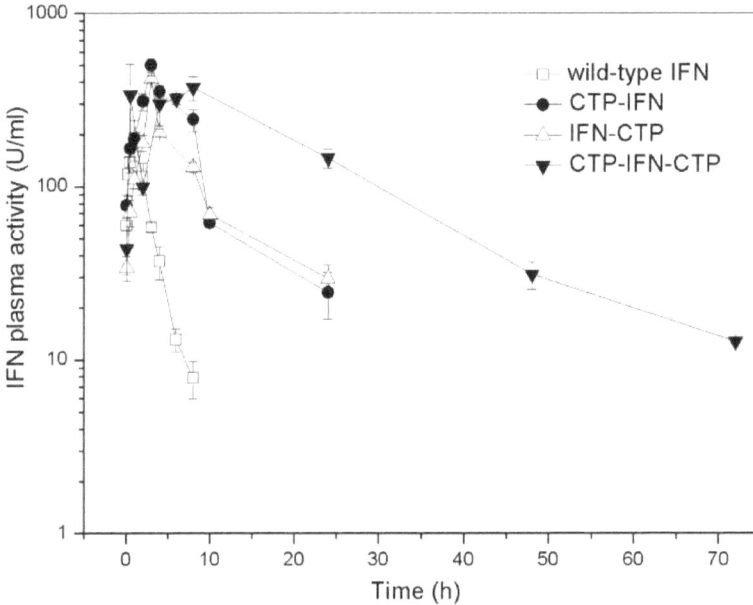

Fig. (7). Pharmacokinetic plasma profile of wild-type IFN, CTP-IFN, IFN-CTP and CTP-IFN-CTP injected subcutaneously in Wistar rats. Data points are the average ± SEM of four animals in each group. Reprinted from Ceaglio *et al.* (2016) [56] with permission from Elsevier.

The results described here exemplify that both N- and O-glycosylation engineering constitute successful strategies to prolong the duration of action of therapeutic proteins, and thus, to improve their *in vivo* efficacy. Besides, as glycans have a profound effect on other properties of glycoproteins, including solubility, stability, protease and thermal resistance, additional advantages could be achieved through these technologies.

CONFLICT OF INTEREST

The authors confirm that they have no conflict of interest to declare for this publication.

ACKNOWLEDGEMENTS

We acknowledge the following Argentine Institutions: Consejo Nacional de

Investigaciones Científicas y Técnicas (CONICET) and Universidad Nacional del Litoral (UNL). Authors also thank Qun Zhou for helping us in proofreading the manuscript. Special thanks to Sebastián Antuña who kindly helped us in the preparation of the figures.

REFERENCES

[1]   Marshall SA, Lazar GA, Chirino AJ, Desjarlais JR. Rational design and engineering of therapeutic proteins. Drug Discov Today 2003; 8(5): 212-21.
        [http://dx.doi.org/10.1016/S1359-6446(03)02610-2] [PMID: 12634013]

[2]   Carter PJ. Introduction to current and future protein therapeutics: a protein engineering perspective. Exp Cell Res 2011; 317(9): 1261-9.
        [http://dx.doi.org/10.1016/j.yexcr.2011.02.013] [PMID: 21371474]

[3]   Szlachcic A, Zakrzewska M, Otlewski J. Longer action means better drug: tuning up protein therapeutics. Biotechnol Adv 2011; 29(4): 436-41.
        [http://dx.doi.org/10.1016/j.biotechadv.2011.03.005] [PMID: 21443940]

[4]   Sheffield WP. Modification of clearance of therapeutic and potentially therapeutic proteins. Curr Drug Targets Cardiovasc Haematol Disord 2001; 1(1): 1-22.
        [http://dx.doi.org/10.2174/1568006013338150] [PMID: 12769661]

[5]   Beals JM, Shanafelt AB. Enhancing exposure of protein therapeutics. Drug Discov Today Technol 2006; 3(1): 87-94.
        [http://dx.doi.org/10.1016/j.ddtec.2006.03.001] [PMID: 24980106]

[6]   Mahmood I, Green MD. Pharmacokinetic and pharmacodynamic considerations in the development of therapeutic proteins. Clin Pharmacokinet 2005; 44(4): 331-47.
        [http://dx.doi.org/10.2165/00003088-200544040-00001] [PMID: 15828849]

[7]   Baumann A. Early development of therapeutic biologicspharmacokinetics. Curr Drug Metab 2006; 7(1): 15-21.
        [http://dx.doi.org/10.2174/138920006774832604] [PMID: 16454690]

[8]   Spiro RG. Protein glycosylation: nature, distribution, enzymatic formation, and disease implications of glycopeptide bonds. Glycobiology 2002; 12(4): 43R-56R.
        [http://dx.doi.org/10.1093/glycob/12.4.43R] [PMID: 12042244]

[9]   Walsh G. Post-translational modifications of protein biopharmaceuticals. Drug Discov Today 2010; 15(17-18): 773-80.
        [http://dx.doi.org/10.1016/j.drudis.2010.06.009] [PMID: 20599624]

[10]   Li H, dAnjou M. Pharmacological significance of glycosylation in therapeutic proteins. Curr Opin Biotechnol 2009; 20(6): 678-84.
        [http://dx.doi.org/10.1016/j.copbio.2009.10.009] [PMID: 19892545]

[11]   Berger M, Kaup M, Blanchard V. Protein glycosylation and its impact on biotechnology. Adv Biochem Eng Biotechnol 2012; 127: 165-85.
        [PMID: 21975953]

[12]   Kontermann RE. Strategies for extended serum half-life of protein therapeutics. Curr Opin Biotechnol 2011; 22(6): 868-76.
[http://dx.doi.org/10.1016/j.copbio.2011.06.012] [PMID: 21862310]

[13]   Varki A. Biological roles of oligosaccharides: all of the theories are correct. Glycobiology 1993; 3(2): 97-130.
[http://dx.doi.org/10.1093/glycob/3.2.97] [PMID: 8490246]

[14]   Goochee CF, Gramer MJ, Andersen DC, Bahr JB, Rasmussen JR. The oligosaccharides of glycoproteins: bioprocess factors affecting oligosaccharide structure and their effect on glycoprotein properties. Biotechnology (N Y) 1991; 9(12): 1347-55.
[http://dx.doi.org/10.1038/nbt1291-1347] [PMID: 1367768]

[15]   Oh-eda M, Hasegawa M, Hattori K, *et al.* O-linked sugar chain of human granulocyte colony-stimulating factor protects it against polymerization and denaturation allowing it to retain its biological activity. J Biol Chem 1990; 265(20): 11432-5.
[PMID: 1694845]

[16]   Goldwasser E, Kung CK, Eliason J. On the mechanism of erythropoietin-induced differentiation. 13. The role of sialic acid in erythropoietin action. J Biol Chem 1974; 249(13): 4202-6.
[PMID: 4368980]

[17]   Wittwer AJ, Howard SC. Glycosylation at Asn-184 inhibits the conversion of single-chain to two-chain tissue-type plasminogen activator by plasmin. Biochemistry 1990; 29(17): 4175-80.
[http://dx.doi.org/10.1021/bi00469a021] [PMID: 2141793]

[18]   Jenkins N, Curling EM. Glycosylation of recombinant proteins: problems and prospects. Enzyme Microb Technol 1994; 16(5): 354-64.
[http://dx.doi.org/10.1016/0141-0229(94)90149-X] [PMID: 7764790]

[19]   Chamorey AL, Magné N, Pivot X, Milano G. Impact of glycosylation on the effect of cytokines. A special focus on oncology. Eur Cytokine Netw 2002; 13(2): 154-60.
[PMID: 12101071]

[20]   Markoff E, Sigel MB, Lacour N, Seavey BK, Friesen HG, Lewis UJ. Glycosylation selectively alters the biological activity of prolactin. Endocrinology 1988; 123(3): 1303-6.
[http://dx.doi.org/10.1210/endo-123-3-1303] [PMID: 3402386]

[21]   Chavin SI, Weidner SM. Blood clotting factor IX. Loss of activity after cleavage of sialic acid residues. J Biol Chem 1984; 259(6): 3387-90.
[PMID: 6323421]

[22]   Lisowska E. The role of glycosylation in protein antigenic properties. Cell Mol Life Sci 2002; 59(3): 445-55.
[http://dx.doi.org/10.1007/s00018-002-8437-3] [PMID: 11964123]

[23]   Revoltella RP, Laricchia-Robbio L, Moscato S, Genua A, Liberati AM. Natural and therapy-induced anti-GM-CSF and anti-G-CSF antibodies in human serum. Leuk Lymphoma 1997; 26 (Suppl. 1): 29-34.
[http://dx.doi.org/10.3109/10428199709058597] [PMID: 9570677]

[24]   Wadhwa M, Meager A, Dilger P, *et al.* Neutralizing antibodies to granulocyte-macrophage colony-

stimulating factor, interleukin-1alpha and interferon-alpha but not other cytokines in human immunoglobulin preparations. Immunology 2000; 99(1): 113-23.
[http://dx.doi.org/10.1046/j.1365-2567.2000.00949.x] [PMID: 10651949]

[25] Oggero M, Kratje R, Etcheverrigaray M. Anti-rhGM-CSF activity in normal human sera. FABICIB Journal 2002; 6: 57-67.

[26] Takeuchi M, Takasaki S, Shimada M, Kobata A. Role of sugar chains in the *in vitro* biological activity of human erythropoietin produced in recombinant Chinese hamster ovary cells. J Biol Chem 1990; 265(21): 12127-30.
[PMID: 2373681]

[27] Baenziger JU. Glycosylation: to what end for the glycoprotein hormones? Endocrinology 1996; 137(5): 1520-2.
[http://dx.doi.org/10.1210/endo.137.5.8612480] [PMID: 8612480]

[28] Morell AG, Gregoriadis G, Scheinberg IH, Hickman J, Ashwell G. The role of sialic acid in determining the survival of glycoproteins in the circulation. J Biol Chem 1971; 246(5): 1461-7.
[PMID: 5545089]

[29] Kawasaki T, Ashwell G. Chemical and physical properties of an hepatic membrane protein that specifically binds asialoglycoproteins. J Biol Chem 1976; 251(5): 1296-302.
[PMID: 1254568]

[30] Cumming DA. Glycosylation of recombinant protein therapeutics: control and functional implications. Glycobiology 1991; 1(2): 115-30.
[http://dx.doi.org/10.1093/glycob/1.2.115] [PMID: 1823155]

[31] Koury MJ. Sugar coating extends half-lives and improves effectiveness of cytokine hormones. Trends Biotechnol 2003; 21(11): 462-4.
[http://dx.doi.org/10.1016/j.tibtech.2003.09.002] [PMID: 14573355]

[32] Wacker M, Linton D, Hitchen PG, *et al.* N-linked glycosylation in *Campylobacter jejuni* and its functional transfer into *E. coli.* Science 2002; 298(5599): 1790-3.
[http://dx.doi.org/10.1126/science.298.5599.1790] [PMID: 12459590]

[33] Pandhal J, Ow SY, Noirel J, Wright PC. Improving N-glycosylation efficiency in Escherichia coli using shotgun proteomics, metabolic network analysis, and selective reaction monitoring. Biotechnol Bioeng 2011; 108(4): 902-12.
[http://dx.doi.org/10.1002/bit.23011] [PMID: 21404263]

[34] Fisher AC, Haitjema CH, Guarino C, *et al.* Production of secretory and extracellular N-linked glycoproteins in *Escherichia coli.* Appl Environ Microbiol 2011; 77(3): 871-81.
[http://dx.doi.org/10.1128/AEM.01901-10] [PMID: 21131519]

[35] Naegeli A, Neupert C, Fan YY, *et al.* Molecular analysis of an alternative N-glycosylation machinery by functional transfer from *Actinobacillus pleuropneumoniae* to *Escherichia coli.* J Biol Chem 2014; 289(4): 2170-9.
[http://dx.doi.org/10.1074/jbc.M113.524462] [PMID: 24275653]

[36] Wang LX, Lomino JV. Emerging technologies for making glycan-defined glycoproteins. ACS Chem Biol 2012; 7(1): 110-22.

[http://dx.doi.org/10.1021/cb200429n] [PMID: 22141574]

[37]   Arico C, Bonnet C, Javaud C. N-glycosylation humanization for production of therapeutic recombinant glycoproteins in *Saccharomyces cerevisiae*. Methods Mol Biol 2013; 988: 45-57.
[http://dx.doi.org/10.1007/978-1-62703-327-5_4] [PMID: 23475713]

[38]   Aumiller JJ, Mabashi-Asazuma H, Hillar A, Shi X, Jarvis DL. A new glycoengineered insect cell line with an inducibly mammalianized protein N-glycosylation pathway. Glycobiology 2012; 22(3): 417-28.
[http://dx.doi.org/10.1093/glycob/cwr160] [PMID: 22042767]

[39]   Mabashi-Asazuma H, Shi X, Geisler C, Kuo CW, Khoo KH, Jarvis DL. Impact of a human CMP-sialic acid transporter on recombinant glycoprotein sialylation in glycoengineered insect cells. Glycobiology 2013; 23(2): 199-210.
[http://dx.doi.org/10.1093/glycob/cws143] [PMID: 23065352]

[40]   Loos A, Steinkellner H. Plant glyco-biotechnology on the way to synthetic biology. Front Plant Sci 2014; 5: 523.
[http://dx.doi.org/10.3389/fpls.2014.00523] [PMID: 25339965]

[41]   Weikert S, Papac D, Briggs J, *et al.* Engineering Chinese hamster ovary cells to maximize sialic acid content of recombinant glycoproteins. Nat Biotechnol 1999; 17(11): 1116-21.
[http://dx.doi.org/10.1038/15104] [PMID: 10545921]

[42]   Butler M, Spearman M. The choice of mammalian cell host and possibilities for glycosylation engineering. Curr Opin Biotechnol 2014; 30: 107-12.
[http://dx.doi.org/10.1016/j.copbio.2014.06.010] [PMID: 25005678]

[43]   Davies J, Jiang L, Pan LZ, LaBarre MJ, Anderson D, Reff M. Expression of GnTIII in a recombinant anti-CD20 CHO production cell line: Expression of antibodies with altered glycoforms leads to an increase in ADCC through higher affinity for FC gamma RIII. Biotechnol Bioeng 2001; 74(4): 288-94.
[http://dx.doi.org/10.1002/bit.1119] [PMID: 11410853]

[44]   Ferrara C, Brünker P, Suter T, Moser S, Püntener U, Umaña P. Modulation of therapeutic antibody effector functions by glycosylation engineering: influence of Golgi enzyme localization domain and co-expression of heterologous beta1, 4-N-acetylglucosaminyltransferase III and Golgi alpha-mannosidase II. Biotechnol Bioeng 2006; 93(5): 851-61.
[http://dx.doi.org/10.1002/bit.20777] [PMID: 16435400]

[45]   Meuris L, Santens F, Elson G, *et al.* GlycoDelete engineering of mammalian cells simplifies N-glycosylation of recombinant proteins. Nat Biotechnol 2014; 32(5): 485-9.
[http://dx.doi.org/10.1038/nbt.2885] [PMID: 24752077]

[46]   Piron R, Santens F, De Paepe A, Depicker A, Callewaert N. Using GlycoDelete to produce proteins lacking plant-specific N-glycan modification in seeds. Nat Biotechnol 2015; 33(11): 1135-7.
[http://dx.doi.org/10.1038/nbt.3359] [PMID: 26544140]

[47]   Elliott S, Lorenzini T, Asher S, *et al.* Enhancement of therapeutic protein *in vivo* activities through glycoengineering. Nat Biotechnol 2003; 21(4): 414-21.
[http://dx.doi.org/10.1038/nbt799] [PMID: 12612588]

[48]   Sinclair AM, Elliott S. Glycoengineering: the effect of glycosylation on the properties of therapeutic

proteins. J Pharm Sci 2005; 94(8): 1626-35.
[http://dx.doi.org/10.1002/jps.20319] [PMID: 15959882]

[49] Egrie JC, Browne JK. Development and characterization of novel erythropoiesis stimulating protein (NESP). Nephrol Dial Transplant 2001; 16 (Suppl. 3): 3-13.
[http://dx.doi.org/10.1093/ndt/16.suppl_3.3] [PMID: 11402085]

[50] Egrie JC, Dwyer E, Browne JK, Hitz A, Lykos MA. Darbepoetin alfa has a longer circulating half-life and greater *in vivo* potency than recombinant human erythropoietin. Exp Hematol 2003; 31(4): 290-9.
[http://dx.doi.org/10.1016/S0301-472X(03)00006-7] [PMID: 12691916]

[51] Fares F, Ganem S, Hajouj T, Agai E. Development of a long-acting erythropoietin by fusing the carboxyl-terminal peptide of human chorionic gonadotropin beta-subunit to the coding sequence of human erythropoietin. Endocrinology 2007; 148(10): 5081-7.
[http://dx.doi.org/10.1210/en.2007-0026] [PMID: 17641000]

[52] Perlman S, van den Hazel B, Christiansen J, *et al.* Glycosylation of an N-terminal extension prolongs the half-life and increases the *in vivo* activity of follicle stimulating hormone. J Clin Endocrinol Metab 2003; 88(7): 3227-35.
[http://dx.doi.org/10.1210/jc.2002-021201] [PMID: 12843169]

[53] Trousdale RK, Yu B, Pollak SV, Husami N, Vidali A, Lustbader JW. Efficacy of native and hyperglycosylated follicle-stimulating hormone analogs for promoting fertility in female mice. Fertil Steril 2009; 91(1): 265-70.
[http://dx.doi.org/10.1016/j.fertnstert.2007.11.013] [PMID: 18249396]

[54] Fares FA, Suganuma N, Nishimori K, LaPolt PS, Hsueh AJ, Boime I. Design of a long-acting follitropin agonist by fusing the C-terminal sequence of the chorionic gonadotropin beta subunit to the follitropin beta subunit. Proc Natl Acad Sci USA 1992; 89(10): 4304-8.
[http://dx.doi.org/10.1073/pnas.89.10.4304] [PMID: 1374895]

[55] Ceaglio N, Etcheverrigaray M, Kratje R, Oggero M. Novel long-lasting interferon alpha derivatives designed by glycoengineering. Biochimie 2008; 90(3): 437-49.
[http://dx.doi.org/10.1016/j.biochi.2007.10.013] [PMID: 18039474]

[56] Ceaglio N, Gugliotta A, Tardivo MB, *et al.* Improvement of *in vitro* stability and pharmacokinetics of hIFN-α by fusing the carboxyl-terminal peptide of hCG β-subunit. J Biotechnol 2016; 221: 13-24.
[http://dx.doi.org/10.1016/j.jbiotec.2016.01.018] [PMID: 26806490]

[57] Samoudi M, Tabandeh F, Minuchehr Z, *et al.* Rational design of hyper-glycosylated interferon beta analogs: a computational strategy for glycoengineering. J Mol Graph Model 2015; 56: 31-42.
[http://dx.doi.org/10.1016/j.jmgm.2014.12.001] [PMID: 25544388]

[58] Song K, Yoon IS, Kim NA, *et al.* Glycoengineering of interferon-β 1a improves its biophysical and pharmacokinetic properties. PLoS One 2014; 9(5): e96967.
[http://dx.doi.org/10.1371/journal.pone.0096967] [PMID: 24858932]

[59] Croxtall JD, McKeage K. Corifollitropin alfa: a review of its use in controlled ovarian stimulation for assisted reproduction. BioDrugs 2011; 25(4): 243-54.
[http://dx.doi.org/10.2165/11206890-000000000-00000] [PMID: 21815699]

[60] Rachmawati H, Poelstra K, Beljaars L. The use of cytokines and modified cytokines as therapeutic

agents: present state and future perspectives. Kerala, India: Research Signpost 2004.

[61]   Weenen C, Peña JE, Pollak SV, *et al.* Long-acting follicle-stimulating hormone analogs containing N-linked glycosylation exhibited increased bioactivity compared with o-linked analogs in female rats. J Clin Endocrinol Metab 2004; 89(10): 5204-12.
[http://dx.doi.org/10.1210/jc.2004-0425] [PMID: 15472227]

[62]   Flintegaard TV, Thygesen P, Rahbek-Nielsen H, *et al.* N-glycosylation increases the circulatory half-life of human growth hormone. Endocrinology 2010; 151(11): 5326-36.
[http://dx.doi.org/10.1210/en.2010-0574] [PMID: 20826563]

[63]   Bolt G, Bjelke JR, Hermit MB, Hansen L, Karpf DM, Kristensen C. Hyperglycosylation prolongs the circulation of coagulation factor IX. J Thromb Haemost 2012; 10(11): 2397-8.
[http://dx.doi.org/10.1111/j.1538-7836.2012.04911.x] [PMID: 22938555]

[64]   Schwarz F, Aebi M. Mechanisms and principles of N-linked protein glycosylation. Curr Opin Struct Biol 2011; 21(5): 576-82.
[http://dx.doi.org/10.1016/j.sbi.2011.08.005] [PMID: 21978957]

[65]   Aebi M. N-linked protein glycosylation in the ER. Biochim Biophys Acta 2013; 1833(11): 2430-7.
[http://dx.doi.org/10.1016/j.bbamcr.2013.04.001] [PMID: 23583305]

[66]   Rudd PM, Dwek RA. Glycosylation: heterogeneity and the 3D structure of proteins. Crit Rev Biochem Mol Biol 1997; 32(1): 1-100.
[http://dx.doi.org/10.3109/10409239709085144] [PMID: 9063619]

[67]   Spahn PN, Lewis NE. Systems glycobiology for glycoengineering. Curr Opin Biotechnol 2014; 30: 218-24.
[http://dx.doi.org/10.1016/j.copbio.2014.08.004] [PMID: 25202878]

[68]   Kornfeld R, Kornfeld S. Assembly of asparagine-linked oligosaccharides. Annu Rev Biochem 1985; 54: 631-64.
[http://dx.doi.org/10.1146/annurev.bi.54.070185.003215] [PMID: 3896128]

[69]   Shakin-Eshleman SH, Spitalnik SL, Kasturi L. The amino acid at the X position of an Asn-X-Ser sequon is an important determinant of N-linked core-glycosylation efficiency. J Biol Chem 1996; 271(11): 6363-6.
[http://dx.doi.org/10.1074/jbc.271.11.6363] [PMID: 8626433]

[70]   Kasturi L, Eshleman JR, Wunner WH, Shakin-Eshleman SH. The hydroxy amino acid in an Asn-X-Ser/Thr sequon can influence N-linked core glycosylation efficiency and the level of expression of a cell surface glycoprotein. J Biol Chem 1995; 270(24): 14756-61.
[http://dx.doi.org/10.1074/jbc.270.24.14756] [PMID: 7782341]

[71]   Valliere-Douglass JF, Eakin CM, Wallace A, *et al.* Glutamine-linked and non-consensus asparagine-linked oligosaccharides present in human recombinant antibodies define novel protein glycosylation motifs. J Biol Chem 2010; 285(21): 16012-22.
[http://dx.doi.org/10.1074/jbc.M109.096412] [PMID: 20233717]

[72]   Gavel Y, von Heijne G. Sequence differences between glycosylated and non-glycosylated Asn-X-Thr/Ser acceptor sites: implications for protein engineering. Protein Eng 1990; 3(5): 433-42.
[http://dx.doi.org/10.1093/protein/3.5.433] [PMID: 2340213]

[73]   Nilsson I, von Heijne G. Glycosylation efficiency of Asn-Xaa-Thr sequons depends both on the distance from the C terminus and on the presence of a downstream transmembrane segment. J Biol Chem 2000; 275(23): 17338-43.
[http://dx.doi.org/10.1074/jbc.M002317200] [PMID: 10748070]

[74]   Elliott S, Chang D, Delorme E, Eris T, Lorenzini T. Structural requirements for additional N-linked carbohydrate on recombinant human erythropoietin. J Biol Chem 2004; 279(16): 16854-62.
[http://dx.doi.org/10.1074/jbc.M311095200] [PMID: 14757769]

[75]   Shental-Bechor D, Levy Y. Folding of glycoproteins: toward understanding the biophysics of the glycosylation code. Curr Opin Struct Biol 2009; 19(5): 524-33.
[http://dx.doi.org/10.1016/j.sbi.2009.07.002] [PMID: 19647993]

[76]   Jones J, Krag SS, Betenbaugh MJ. Controlling N-linked glycan site occupancy. Biochim Biophys Acta 2005; 1726(2): 121-37.
[http://dx.doi.org/10.1016/j.bbagen.2005.07.003] [PMID: 16126345]

[77]   Hossler P, Khattak SF, Li ZJ. Optimal and consistent protein glycosylation in mammalian cell culture. Glycobiology 2009; 19(9): 936-49.
[http://dx.doi.org/10.1093/glycob/cwp079] [PMID: 19494347]

[78]   Gupta R, Brunak S. Prediction of glycosylation across the human proteome and the correlation to protein function. Pac Symp Biocomput 2002; 310-22.
[PMID: 11928486]

[79]   Chauhan JS, Rao A, Raghava GP. In silico platform for prediction of N-, O- and C-glycosites in eukaryotic protein sequences. PLoS One 2013; 8(6): e67008.
[http://dx.doi.org/10.1371/journal.pone.0067008] [PMID: 23840574]

[80]   Chuang GY, Boyington JC, Joyce MG, *et al.* Computational prediction of N-linked glycosylation incorporating structural properties and patterns. Bioinformatics 2012; 28(17): 2249-55.
[http://dx.doi.org/10.1093/bioinformatics/bts426] [PMID: 22782545]

[81]   Ahmad S, Gromiha M, Fawareh H, Sarai A. ASAView: database and tool for solvent accessibility representation in proteins. BMC Bioinformatics 2004; 5: 51.
[http://dx.doi.org/10.1186/1471-2105-5-51] [PMID: 15119964]

[82]   Jensen PH, Kolarich D, Packer NH. Mucin-type O-glycosylationputting the pieces together. FEBS J 2010; 277(1): 81-94.
[http://dx.doi.org/10.1111/j.1742-4658.2009.07429.x] [PMID: 19919547]

[83]   Halim A, Brinkmalm G, Rüetschi U, *et al.* Site-specific characterization of threonine, serine, and tyrosine glycosylations of amyloid precursor protein/amyloid beta-peptides in human cerebrospinal fluid. Proc Natl Acad Sci USA 2011; 108(29): 11848-53.
[http://dx.doi.org/10.1073/pnas.1102664108] [PMID: 21712440]

[84]   Smythe C, Caudwell FB, Ferguson M, Cohen P. Isolation and structural analysis of a peptide containing the novel tyrosyl-glucose linkage in glycogenin. EMBO J 1988; 7(9): 2681-6.
[PMID: 3181138]

[85]   Bennett EP, Mandel U, Clausen H, Gerken TA, Fritz TA, Tabak LA. Control of mucin-type O-glycosylation: a classification of the polypeptide GalNAc-transferase gene family. Glycobiology 2012;

22(6): 736-56.
[http://dx.doi.org/10.1093/glycob/cwr182] [PMID: 22183981]

[86]   Spiro RG. Characterization and quantitative determination of the hydroxylysine-linked carbohydrate units of several collagens. J Biol Chem 1969; 244(4): 602-12.
[PMID: 4305879]

[87]   Schjoldager KT, Clausen H. Site-specific protein O-glycosylation modulates proprotein processing - deciphering specific functions of the large polypeptide GalNAc-transferase gene family. Biochim Biophys Acta 2012; 1820(12): 2079-94.
[http://dx.doi.org/10.1016/j.bbagen.2012.09.014] [PMID: 23022508]

[88]   Van den Steen P, Rudd PM, Dwek RA, Opdenakker G. Concepts and principles of O-linked glycosylation. Crit Rev Biochem Mol Biol 1998; 33(3): 151-208.
[http://dx.doi.org/10.1080/10409239891204198] [PMID: 9673446]

[89]   Yang Z, Halim A, Narimatsu Y, *et al.* The GalNAc-type O-Glycoproteome of CHO cells characterized by the SimpleCell strategy. Mol Cell Proteomics 2014; 13(12): 3224-35.
[http://dx.doi.org/10.1074/mcp.M114.041541] [PMID: 25092905]

[90]   Steentoft C, Vakhrushev SY, Joshi HJ, *et al.* Precision mapping of the human O-GalNAc glycoproteome through SimpleCell technology. EMBO J 2013; 32(10): 1478-88.
[http://dx.doi.org/10.1038/emboj.2013.79] [PMID: 23584533]

[91]   Kong Y, Joshi HJ, Schjoldager KT, *et al.* Probing polypeptide GalNAc-transferase isoform substrate specificities by *in vitro* analysis. Glycobiology 2015; 25(1): 55-65.
[http://dx.doi.org/10.1093/glycob/cwu089] [PMID: 25155433]

[92]   Coltart DM, Royyuru AK, Williams LJ, *et al.* Principles of mucin architecture: structural studies on synthetic glycopeptides bearing clustered mono-, di-, tri-, and hexasaccharide glycodomains. J Am Chem Soc 2002; 124(33): 9833-44.
[http://dx.doi.org/10.1021/ja020208f] [PMID: 12175243]

[93]   Julenius K, Mølgaard A, Gupta R, Brunak S. Prediction, conservation analysis, and structural characterization of mammalian mucin-type O-glycosylation sites. Glycobiology 2005; 15(2): 153-64.
[http://dx.doi.org/10.1093/glycob/cwh151] [PMID: 15385431]

[94]   Mrázek H, Weignerová L, Bojarová P, Novák P, Vaněk O, Bezouška K. Carbohydrate synthesis and biosynthesis technologies for cracking of the glycan code: recent advances. Biotechnol Adv 2013; 31(1): 17-37.
[http://dx.doi.org/10.1016/j.biotechadv.2012.03.008] [PMID: 22484115]

[95]   Pierce JG, Parsons TF. Glycoprotein hormones: structure and function. Annu Rev Biochem 1981; 50: 465-95.
[http://dx.doi.org/10.1146/annurev.bi.50.070181.002341] [PMID: 6267989]

[96]   Mazola Y, Chinea G, Musacchio A. Glycosylation and Bioinformatics: current status for glycosylation prediction tools. Biotecnol Apl 2011; 28: 6-12.

[97]   Chen YZ, Tang YR, Sheng ZY, Zhang Z. Prediction of mucin-type O-glycosylation sites in mammalian proteins using the composition of k-spaced amino acid pairs. BMC Bioinformatics 2008; 9: 101.

[http://dx.doi.org/10.1186/1471-2105-9-101] [PMID: 18282281]

[98] Leung M-Y, Cardenas GA, Almeida IC, Gerken TA. Isoform Specific O-Glycosylation Prediction (ISOGlyP)Version 12 , 2014 [Accessed 10/31/2014]; Available at http://isoglyp.utep.edu.

[99] de Medeiros SF, Norman RJ. Human choriogonadotrophin protein core and sugar branches heterogeneity: basic and clinical insights. Hum Reprod Update 2009; 15(1): 69-95.
[http://dx.doi.org/10.1093/humupd/dmn036] [PMID: 18945715]

[100] Cole LA, Khanlian SA. Hyperglycosylated hCG: a variant with separate biological functions to regular hCG. Mol Cell Endocrinol 2007; 260-262: 228-36.
[http://dx.doi.org/10.1016/j.mce.2006.03.047] [PMID: 17081684]

[101] Matzuk MM, Hsueh AJ, Lapolt P, Tsafriri A, Keene JL, Boime I. The biological role of the carboxyl-terminal extension of human chorionic gonadotropin [corrected] beta-subunit. Endocrinology 1990; 126(1): 376-83.
[http://dx.doi.org/10.1210/endo-126-1-376] [PMID: 2293995]

[102] LaPolt PS, Nishimori K, Fares FA, Perlas E, Boime I, Hsueh AJ. Enhanced stimulation of follicle maturation and ovulatory potential by long acting follicle-stimulating hormone agonists with extended carboxyl-terminal peptides. Endocrinology 1992; 131(6): 2514-20.
[PMID: 1446593]

[103] Joshi L, Murata Y, Wondisford FE, Szkudlinski MW, Desai R, Weintraub BD. Recombinant thyrotropin containing a beta-subunit chimera with the human chorionic gonadotropin-beta carboxy-terminus is biologically active, with a prolonged plasma half-life: role of carbohydrate in bioactivity and metabolic clearance. Endocrinology 1995; 136(9): 3839-48.
[PMID: 7544273]

[104] Furuhashi M, Shikone T, Fares FA, Sugahara T, Hsueh AJ, Boime I. Fusing the carboxy-terminal peptide of the chorionic gonadotropin (CG) beta-subunit to the common alpha-subunit: retention of O-linked glycosylation and enhanced *in vivo* bioactivity of chimeric human CG. Mol Endocrinol 1995; 9(1): 54-63.
[PMID: 7539107]

[105] Fares F, Guy R, Bar-Ilan A, Felikman Y, Fima E. Designing a long-acting human growth hormone (hGH) by fusing the carboxyl-terminal peptide of human chorionic gonadotropin beta-subunit to the coding sequence of hGH. Endocrinology 2010; 151(9): 4410-7.
[http://dx.doi.org/10.1210/en.2009-1431] [PMID: 20660071]

[106] Fares F, Havron A, Fima E. Designing a long acting erythropoietin by fusing three carboxyl-terminal peptides of human chorionic gonadotropin β subunit to the N-terminal and C-terminal coding sequence. Int J Cell Biol 2011; 2011: 275063.
[http://dx.doi.org/10.1155/2011/275063] [PMID: 21869890]

[107] Cole LA. Hyperglycosylated hCG, a review. Placenta 2010; 31(8): 653-64.
[http://dx.doi.org/10.1016/j.placenta.2010.06.005] [PMID: 20619452]

[108] Isaacs A, Lindenmann J. Virus interference. I. The interferon. Proc R Soc Lond B Biol Sci 1957; 147(927): 258-67.
[http://dx.doi.org/10.1098/rspb.1957.0048] [PMID: 13465720]

[109] Gresser I, Bourali C, Lévy JP, Fontaine-Brouty-Boyé D, Thomas MT. Increased survival in mice inoculated with tumor cells and treated with interferon preparations. Proc Natl Acad Sci USA 1969; 63(1): 51-7.
[http://dx.doi.org/10.1073/pnas.63.1.51] [PMID: 5257966]

[110] Neumann AU, Lam NP, Dahari H, *et al.* Hepatitis C viral dynamics *in vivo* and the antiviral efficacy of interferon-alpha therapy. Science 1998; 282(5386): 103-7.
[http://dx.doi.org/10.1126/science.282.5386.103] [PMID: 9756471]

[111] Goldstein D, Laszlo J. Interferon therapy in cancer: from imaginon to interferon. Cancer Res 1986; 46(9): 4315-29.
[PMID: 2425950]

[112] Borden EC, Lindner D, Dreicer R, Hussein M, Peereboom D. Second-generation interferons for cancer: clinical targets. Semin Cancer Biol 2000; 10(2): 125-44.
[http://dx.doi.org/10.1006/scbi.2000.0315] [PMID: 10936063]

[113] Gutterman JU. Cytokine therapeutics: lessons from interferon alpha. Proc Natl Acad Sci USA 1994; 91(4): 1198-205.
[http://dx.doi.org/10.1073/pnas.91.4.1198] [PMID: 8108387]

[114] Ceaglio N, Etcheverrigaray M, Kratje R, Oggero M. Influence of carbohydrates on the stability and structure of a hyperglycosylated human interferon alpha mutein. Biochimie 2010; 92(8): 971-8.
[http://dx.doi.org/10.1016/j.biochi.2010.04.004] [PMID: 20403411]

[115] Ceaglio N, Etcheverrigaray M, Conradt HS, Grammel N, Kratje R, Oggero M. Highly glycosylated human alpha interferon: An insight into a new therapeutic candidate. J Biotechnol 2010; 146(1-2): 74-83.
[http://dx.doi.org/10.1016/j.jbiotec.2009.12.020] [PMID: 20067809]

SUBJECT INDEX

A

Activity 6, 7, 39, 46, 47, 49, 61, 91, 102, 111, 113, 115, 119, 123, 130, 177, 184, 186, 191, 195, 197
 antiproliferative 195, 197
 antithrombin 46
 broad-spectrum 39, 49
ADCC activity 26, 142, 152, 155
 enhanced 142, 152
ADCC enhancement 155, 158
ADCC enhancement in non-fucosylated antibodies 157
Addition of N-glycans 180, 182
Aglycosylated Fc 137
All-trans retinoic acid (ATRA) 77
Amino acid residues 182, 183, 184, 185
Aminodextran 78
Amino group of OVA 121, 122
Analysis, theoretical 183, 192, 193
Antibodies 3, 4, 8, 10, 11, 13, 18, 19, 20, 22, 23, 24, 26, 27, 28, 60, 65, 117, 118, 133, 134, 141, 143, 152, 153, 154, 155, 156, 158, 159, 160, 161, 162, 163, 164, 175, 176, 178, 182
 anti-CCR4 159
 anti-CD22 163
 anti-CD30 162
 anti-EPO 176
 anti-glycan 27
 anti-TF 26
 bispecific 23
 chimeric 26
 effective 23
 glycan-containing 155
 glycoengineered 164
 last-mentioned 24
 natural 176
 next-generation 134
 non-conjugated 118
 nonfucosylated 158, 159
 producing 158
 radiolabeling 163
 recombinant 141, 158, 182
 sialylated 152

Antibodies effector functions 23
Antibodies features, intrinsic 23
Antibodies isotypes 10
Antibodies production 22
Antibody conjugation 152, 153, 161
Antibody-dependent 22, 130, 133, 141, 152, 154, 155, 156, 157, 159
 cellular cytotoxicity (ADCC) 22, 130, 141, 152, 154, 155, 156, 157, 159
 cellular phagocytosis (ADCP) 133, 141
Antibody-drug 23, 153, 161, 162, 163
 conjugates (ADC) 23, 153, 161, 162, 163
 conjugation 161, 162
Antibody function 141, 152, 155, 156
Antibody glycoengineering 152, 155, 162, 163
Antibody glycosylation 154, 155
Antibody IgG 153, 154
Antibody molecule 162
Anticancer drug 102, 106, 161, 162, 163, 164
Anticancer vaccines 11, 12, 13, 17, 18, 21
 carbohydrate-based 18
Anti-carbohydrate antibodies 23
Antigen binding 142, 143
 fragment 142
Antigen-binding affinity 142, 143
Antigens 4, 5, 6, 7, 8, 10, 11, 12, 13, 14, 16, 17, 18, 19, 21, 22, 23, 24, 26, 28, 121, 124, 153, 155, 159, 162
 associated 23, 28
 foreign 121
 glycan 24
 native 19, 21
 pan-carcinoma 8
 target 12, 13, 21, 22
Anti-inflammatory activity 152, 156, 157, 159, 160, 164
Anti-proteinase 154
Anti-STn antibodies 7, 9, 14
Anti-Tn antibodies 8
Anti-TNF-α 102, 117, 118
Antiviral protein OAS 116
Antiviral target 45
Asialoglycoprotein 137
Asialoglycoprotein receptor 65, 67, 81, 83, 85, 91, 137, 177

Ovine submaxillary mucin (OSM) 25

P

Paclitaxel 64, 102, 106, 107, 109, 110
P-aminophenyl-α-D-mannopyranoside 89
P-aminophenyl glycoside 83, 84
PEG amine 72
PEG derivatives 62
PEG-Intron 64, 116
PEG polymers 75, 77
PEG-sialic acid 60, 66, 67
PEGylated carbohydrates 60, 63
PEGylated glycoproteins 71
PEGylated nanoparticles 60
PEGylated polysaccharides 60, 75
PEGylated proteins 63, 138
PEGylation 61, 62, 63, 64, 66, 67, 68, 69, 70,
       71, 72, 75, 76, 77, 83, 91, 131, 138
PEGylation method for proteins 66
PEGylation of native glycoproteins 68
PEGylation of proteins 60
Pentavalent construct 18
Peptide backbone 66, 179, 187, 189
Peptide fusion 173
Peptides 9, 10, 46, 102, 103, 106, 112, 113, 114,
       115, 139, 185, 186, 189, 190
   carboxyl-terminal 189, 190
   conjugated 114, 115
   derived 9, 10
   free 114
Peptide substrate specificities 187, 189
Periodate 68, 162
Permethylation 135, 136
P-Nitrophenyl carbonate ester 64
Polycarboxylates 46
Polyethylene glycol 60, 88, 103, 138
Polyethylenimine 85, 86
Polymer conjugate, tumor-targetable 121
Polymer-drug conjugates 103
Polymeric backbone 115, 116, 117, 121
Polymer-protein/peptide conjugates 103
Polymer therapeutics 103
Polypeptide   N-
       acetylgalactosaminyltransferases 6, 187
Post-translational modifications (PTMs) 136,
       172, 174

Prostate cancer 16, 18
Proteases 112, 135, 155, 172, 198
Protein A-binding sites 137
Protein activity 116
Protein aggregation 162, 174
   showed low 162
Protein amino groups 63
Protein-based biotherapeutics 172, 173
Protein bioactivity 121, 180
Protein carrier 106
Protein concentrations, higher 131
Protein conjugation, selective N-terminal 118
Protein context 176
Protein coupling 117
Protein drugs 131
Protein extraction 117
Protein folding, lectins support 132
Protein folding events 187
Protein fragments 121
Protein glycoengineering 189
Protein glycoforms 142
Protein glycosylation 177, 187
Protein-protein interaction, strong 160
Proteins 8, 19, 20, 25, 40, 41, 43, 44, 60, 61, 62,
       63, 64, 65, 66, 67, 68, 70, 91, 102, 103,
       106, 109, 113, 115, 116, 119, 120, 121,
       131, 132, 133, 134, 135, 136, 137, 138,
       139, 140, 141, 160, 162, 164, 172, 173,
       174, 175, 176, 177, 178, 179, 180, 181,
       182, 183, 184, 185, 186, 188, 189, 190,
       191, 194, 195, 196
   administered 173
   altered 66
   anti-fibroblast activation 162
   carrying OSM 25
   conjugated 115
   emblematic 173
   extracellular matrix 132
   fiber shaft 43
   fusion 131
   glycan-bearing 177
   human 140, 172, 174, 176, 179
   human homolog 160
   hyper-N-glycosylated 183
   intracellular 65
   low molecular weight 138
   lymphocyte-associated 140
   mammalian 189

www.ingramcontent.com/pod-product-compliance
Lightning Source LLC
Chambersburg PA
CBHW050837220326
41598CB00006B/387